John Perry

Practical Mechanics

by John Perry

John Perry

Practical Mechanics
by John Perry

ISBN/EAN: 9783337399634

Printed in Europe, USA, Canada, Australia, Japan

Cover: Foto ©berggeist007 / pixelio.de

More available books at **www.hansebooks.com**

PRACTICAL MECHANICS.

BY

JOHN PERRY, M.E.,

Professor of Mechanical Engineering and Applied Mathematics at the City and Guilds of London Technical College, Finsbury.

WITH NUMEROUS ILLUSTRATIONS.

CASSELL, PETTER, GALPIN & CO.
LONDON, PARIS & NEW YORK.

1883.

PREFACE.

This is an attempt to put before non-mathematical readers a *method* of studying mechanics. The student will not benefit much by merely reading the book, nor will he benefit much even if he supplements his reading by listening to lectures on mechanics; but I believe that if by means of lectures he obtains a thorough comprehension of the book, and then makes common-sense experiments with the simple apparatus which is to be found even in the poorest laboratories, but which has hitherto been used merely to illustrate lectures; if, in fact, he uses this book to study mechanics in the manner herein recommended, he will gain in a short time such a working knowledge of the subject as will well repay his labour. I am quite sure also that the mental training acquired in this way is of a kind not inferior to that the belief in which retains in our schools the study of ancient classics and Euclid.

The principle of my method is one which I have tested in practice during the last twelve years, in an English Public School, at the Imperial College of Engineering in Japan, and in other places. It is simply the practical recognition of the fact that all experimenting must be quantitative. It may exercise the wonder of a child, it may please his senses to see certain well-known lecture illustrations which are but little better than tricks, and possibly for young children there may be great instruction in such exhibitions, but they contain no instruction for thinking men who can obtain sufficient amusement elsewhere. Our subject must be studied through quantitative experiments, and when this method of study is adopted it is but of little consequence at what part of the subject the student begins, so long as he begins from his own natural standpoint, the standpoint given him by all his experience.

The primary fact in technical education not yet sufficiently recognised is this—that illiterate men often acquire and possess a useful knowledge of the principles underlying their trade. But the theory usually acted upon is that a man must be quite ignorant of the principles of his trade unless he has been led up to them through weary years' study of the elementary principles of science. On the contrary, the wise apprentice sees for himself that some illiterate journeyman gets good wages, is well thought of by his master, is able to do his work better, perhaps, than any other man in the shop, can be trusted in emergencies, and has a confidence in himself which experience justifies; and the apprentice feels that although the journeyman might be a better workman if he knew the elementary principles of science, still, somehow or other, he has gained an exact knowledge of those more complicated laws with which he has to deal in his trade.

It is good for a man to know the well-established elementary principles of science, to which all complicated laws can be reduced; this enables him to compare his own experience with that of all other people, and enables him to make better use of his own observations in the future.

In giving this knowledge, however, the usual plan of operations is to act on the assumption that the man knows nothing, because he did not begin his previous study with Euclid's axioms, and to teach him the elementary principles as schoolboys of no experience are taught. Now, the standpoint of an experienced workman in the nineteenth century is very different from that of an Alexandrian philosopher or of an English schoolboy, and many men who energetically begin the study of Euclid give it up after a year or two in disgust, because at the end they have only arrived at results which they knew experimentally long ago.

I am inclined to believe that if, instead of forcing the workman to study like a schoolboy, we were to teach the boy as if he had already acquired some of the experience of the workman, and made it our business to give him this experience, we should do better than at present. That is, let the boy work in wood and metal, let him gain experience in the use of machines, let him use drawing instruments and scales, and you put him in a condition to understand and appreciate the truth of the fundamental laws of nature, such a condition as boys usually arrive at only after years of study. It is true he may regard the 47th proposition of the First Book of Euclid as axiomatic; he may think the important propositions in the Sixth Book as easy to believe in as those of the First; he may have greater doubts as to the universal truth of these propositions than mathematicians usually have; but it is possible that these evils are not unmixed with good.

The readers of this book are supposed to have some previous knowledge of the behaviour of materials and machinery. My aim is to give the student such a training as will cause him to think exactly, to give him a method of studying whatever phenomena happen to come before his eyes. Phenomena which, when carelessly considered in the light of elementary principles, appear to follow complicated laws are often found to follow approximately simple laws of their own. A man who knows these roughly correct laws is in a good position for learning the fundamental principles of mechanics; but his teacher must try to view the subject from his student's standpoint, else he cannot take advantage of the fact that his pupil may already possess an excellent foundation on which a superstructure of knowledge may be built. I believe that the most illiterate men may be rapidly taught practical mechanics if we take the right way to teach them—approach the subject from their point of view rather than compel them to approach it from ours.

In a book which is to be used as a general class book by boys and men it is impossible to assume that the reader has an extensive previous practical acquaintance with natural phenomena; but it will be seen that some such past experience is assumed, much more than is usually ascribed to the ordinary student of mechanics. Moreover, he is credited with the possession of common sense, and with the feeling that all human knowledge, instinctive and rational, is the result of experience.

There is much in this book which may seem new to the reader, but inasmuch as I have been, and still am, a student, and as no man can go through life without gathering to himself and regarding as his own many notions of other men, it is probable that there is nothing here, either in the matter or method, which is wholly my own. My partnership in scientific work with Professor Ayrton during the last seven years would in itself preclude any thought of such ownership. For the treatment of some parts of the subject I know that I am indebted to my recollection of the lectures of Professor James Thomson, delivered when I was one of his students fourteen years ago. How much is owing to Sir William Thomson, to Thomson and Tait, and to Professor Ball it is quite impossible to say.

For the careful correction of proofs I have to thank my assistant, Mr. William Robinson, M.E., who has taken as much pains to eliminate numerical errors, puzzling sentences, and crudities of language as if the book had been his own.

I would recommend the student to omit from a first reading the part of the text contained in small type.

This comprises a more detailed and generally more difficult treatment of the matters referred to in the other parts of the text, and it will therefore be best understood by one who has already grasped the general principles of the book.

JOHN PERRY.

71, *Queen Street*,
    London, E.C.

CONTENTS.

CHAPTER I.
INTRODUCTORY.
PAGE

1. What I expect the Reader to Know—2. Equilibrium of Forces—The Triangle of Forces and the Polygon of Forces—3. Our one Theory insufficient—Friction, a Passive Force. 1

CHAPTER II.
FRICTION IN MACHINES.

4. Law of Work—Velocity Ratio—5. Effect of Friction—6. Use of Squared Paper—7. Law connecting Two Variable Things found by means of Squared Paper—8. Law of Friction—9. Force of Friction—10. Loss of Energy due to Friction—11. Friction at Bearings of Shafts—12. Friction in Parallel Motion—13. Friction in Quick-Moving Shafts—14. Mechanical Advantage—15. Rate of doing Work—16. Economical Efficiency—Law for a Machine 5

CHAPTER III.
MACHINES—SPECIAL CASES.

17. Blocks and Tackle—Law of Work—18. Inclined Plane—19. The Screw—20. Differential Pulley-block—21. Wheel and Axle—22. Equilibrium in One Position—23. Body turning about an Axis—Law of Moments—24. The Lever—Weighted Safety-valve—Weighbridge—25. Hydraulic Press 17

CHAPTER IV.
MACHINERY IN GENERAL.

26. Mechanism—27. Velocity Ratio—Toothed Gearing—28. Shapes of Teeth—Worm and Worm-wheel—29. Skeleton Drawings—Crank and Connecting Rod—Pure Harmonic Motion—Eccentric—30. How a Shaft transmits Power—Couplings—31. Belts—Friction between Cord and Pulley—32. Transmission and Absorption Dynamometers 26

CHAPTER V.
FLY-WHEELS.

33. Kinetic Energy—Potential Energy—34. Energy Indestructible—Simple Pendulum—Law for Kinetic Energy stored up in a Moving body—35. Test of the Law—Atwood's Machine—36. Energy in a

Rotating Body—37. Energy in a Wheel when Rotating, experimentally Found—38. Friction never is Negligible—The M of a Wheel—39. Counting the Number of Revolutions—40. Method to be Employed—41. The M of a Wheel—Its Value in different Cases—42. Formula for Energy stored up in a Rotating Body—43. Examples of Energy stored up in Wheels—44. Steadiness of Machines—Examples of Similar Wheels—45. Total Kinetic Energy stored up in any Machine—46. Facts useful to Know 38

CHAPTER VI.

EXTENSION AND COMPRESSION.

47. How a Pull is Exerted—48. Strain—Extension of a Wire—49. Stress—Extension of a Tie-rod—50. Shortening of a Strut—Struts and Ties—51. A Short Strut—Young's Modulus of Elasticity—52. Permanent Set—Limits of Elasticity—53. Nature of Strain—54. Illustration of the Nature of Strain—55. Modulus of Elasticity of Bulk—56. Lateral Contraction—57. Tensile and Compressive Strength—Examples—58. Strength of Pipes and Boilers—59. Tendency of a Boiler to Burst Laterally and Longitudinally—Stress on a Spherical Boiler 51

CHAPTER VII.

PECULIAR BEHAVIOUR OF MATERIALS.

60. Elastic Strength affected by State of Strain—Killing of Wire—Annealing—61. Fatigue of Materials—62. Effect of Load Suddenly Applied—63. Strain Energy—64. Effect of Tensile and Compressive Stresses often Repeated—65. Tempering of Steel—Strengthening of Metal Wires and Cast Iron—66. Curious Properties of Materials which Workmen know of 62

CHAPTER VIII.

MATERIALS.

67 Importance of the Study of Scientific Principles—68. Stones—Structural Characters of Rocks—Preservation of Stone—An Artificial Stone—69. Bricks—70. Lime—Cement—Mortar and Concrete—71. Pressure of Earth against a Wall—72. Pressure of Water—Total Energy of Water—Flow of Water in Pipes and Pumps—Frictional Loss—73. Discharge from Orifices and Pipes—74. Timber—Structure of Timber—Firwoods—Larch—Cedar—Oak—Teak—Mahogany—Ash—Elm—Beech—Time for Felling Timber—Effect of Seasoning—Preservation of Timber—75. Glass—76. Cast Iron—77. Patterns and Moulding—78. The Cooling of Castings—79. Wrought Iron—80. Steel—81. Copper—82. Alloys of Copper—Brass—Munts Metal—Bronze and Gun-Metal—Phosphor Bronze 70

CHAPTER IX.

SHEAR AND TWIST.

83. Shear Stress and Shear Strain—84. Deflection due to Shearing—85 Breaking Shear Stress and Working Shear Stress—86. Punching and Shearing—87. Pure Shear Strain—88. Nature of Shear Strain

CONTENTS. xiii

PAGE

—89. Investigation of the Relation between Shear Stress and Shear Strain—Modulus of Rigidity—90. General Results—91. Twisting—Angle of Twist—Twisting Moment—Rule used by Engineers—92. Investigation of the Twist in a Round Shaft—93. Strength of Shafts—94. Effects of a Twisting Couple in Shafts other than Circular—95. Greatest Distortion at that part of the Surface nearest the Axis—Re-entrant Edge or Angle, a Weak Point—Effects of Twisting on Different Sections—96. Elastic Strength varies with the State of Strain—Two Limits of Elasticity for Loads Twisting in Opposite Directions—97. Stiffness necessary as well as Strength . . . 86

CHAPTER X.

BENDING.

98. State of Strain in a Loaded Beam—99. Distribution of Strain—100. Bending Moment and Shearing Force—101. Case of Pure Bending—102. Neutral Axis passes through the Centre of Gravity—103. Moment of Inertia and Strength Modulus—104. Stress at any part of Section—Best Section for a particular Material—105. Radius of Curvature of Bent Beam—Investigation and Example—106. Elastic Curve and How to Draw it—107. Hydrostatic Arch—108. Alteration of Cross Section of Elastic Strip due to Bending—108 a. Relation between Bending and Twisting 101

CHAPTER XI.

BEAMS.

109. Distribution of the Loads on a Beam—110. Methods of Supporting Beams—111. Supporting Forces at Ends of Beam—112. Diagrams of Bending Moment—113. Shearing Force in Beams and Girders—114. Flanges made to Resist Bending Moment—115. Beam of Uniform Strength—116. Uniform Rectangular Beam Loaded in Various Ways—Rule for Breaking Load—117. Factors of Safety—118. Strength of Beams—119. Curvature of a Loaded Beam—120. Deflection of a Beam—Formulæ and Examples—121. Experiments on Deflection of Beams—122. Rule and Example—123. Stiffness of Beams—124. How Strength and Stiffness vary with the Dimensions—125. Results of Experiments with Testing Machine 112

CHAPTER XII.

BENDING AND CRUSHING.

126. Stress over a Section—127. Resultant Compressive and Tensile Stresses—128. Struts and Pillars—129. Strength of Struts—130. The Teeth of Wheels—131. Strength of Flat Plates—132. Similar Structures Similarly Loaded 128

CHAPTER XIII.

GRAPHICAL STATICS.

133, 134. Methods of Calculation—135. Forces acting at a Point—The Force Polygon—136. The Link Polygon—137. Interpretation of Force and Link Polygon—138. Propositions to be Proved by Actual

Drawing—139. Forces acting on a Ladder—140. Graphical Determination of the Centre of Gravity and Moment of Inertia—141. Formulæ for Centre of Gravity and Moment of Inertia of any Area—142. Poinsot's Theorem regarding the Moment of Inertia . . 135

CHAPTER XIV.

EXAMPLES IN GRAPHICAL STATICS.

143. Diagrams of Bending Moment—144. Shape of a Loaded Beam—145. Hinged Structures—145a. Straight-line Figures—146. Reciprocal Figures—147. Calculation of the Stresses in any Piece of a Hinged Structure—148. Method of Determining the Stresses in a Roof-principal or other Structure—149. Roofs—Practical Example of a Roof—Examples of Stress Diagrams—150. Stiff Joints in a Structure—151. Stresses at any Section of a Loaded Structure calculated by Method of Moments 144

CHAPTER XV.

SUSPENSION BRIDGES, ARCHES, AND BUTTRESSES.

152. Loaded Links—153, 154. Loaded Chain—Curve in which it Hangs and the Pull in Any Part—155. Arched Rib—156. Distribution of the Load on an Arch—Link Polygon must lie Within Middle Third of Arch Ring—157. Fuller's Method of Drawing Link Polygon—158. Iron Arches—159. Buttresses 150

CHAPTER XVI.

SPIRAL SPRINGS.

160. Bending of a Flat Spiral Spring—Change of Curvature—161. Forces acting on Arbor—Bending Moment—162. Couple Proportional to Angle of Winding—163. Experiments on Mainsprings—164. Investigation of Forces acting on a Cylindric Spiral Spring—165. Relation between Extension of Spring and Twisting of Wire—166. Exercise—167. Investigation of Stiffness of Spring—168. Energy stored up in a Spiral Spring—169. Readings of Formulæ—170. Investigation of Elastic Strength—171. Spiral Spring under the Action of a Weight 167

CHAPTER XVII.

PERIODIC MOTION.

172. Periodic Motion and Periodic Time—173. Pure Harmonic Motion—174. Average Velocity Represented—175. Acceleration at any Place—Rule for Finding the Periodic Time—176. Example—Heavy Ball carried by Spiral Spring—177. The Simple Pendulum—Time of an Oscillation—178. Example—Strip of Steel—179. Liquid in Bent Glass Tube—180. Continual Conversion of Energy—Office of Mainspring—Escapements 179

CHAPTER XVIII.

OTHER EXAMPLES OF PERIODIC MOTION.

181. Combination of Pure Harmonic Motions—Thomson's Tide Predicter—182. Blackburn's Pendulum—183. Representation of Motion on Paper—Experimental Illustrations—184. Periodic Rotational

CONTENTS.

PAGE

Motion—Balance of Watch—185. Compound Pendulum—Equivalent Simple Pendulum—Kater's Pendulum—Determination of G—Axes of Oscillation and Suspension are Interchangeable—186. Examples—Experiments on the Twisting Moments of Wires and Flat Spiral Springs—Determination of Viscosity—Friction in Fluids—Bifilar Suspension—Time of Vibration of a Magnet—187. Stilling of Vibrations — Representation of Damped Vibrations — Relative Viscosities of Fluids 188

CHAPTER XIX.

THE EFFECT OF A BLOW.

189. Average Force of an Impact—189. Example—Pile Driver—190. Total Momentum unaltered by Impact—191. Examples—Recoil of Cannon—The Steamship *Waterwitch*—Principle of Barker's Millwork, and Hero's Steam-Engine—192. Mean Pressure during Impact—193. Communication of Momentum through a Liquid—194. Impact of Free Bodies—Storage of Energy during Impact—195. Communication of Strain Energy dependent on Shape of Body—196. Example—Candle Fired through Board—197. Earthquake Effects—198. Examples Mentioned—199. Quasi-Rigidity produced by Rapid Motion—200. Motion produced by a Blow—Centre of Percussion—201. Ballistic Pendulum 206

CHAPTER XX.

THE BALANCING OF MACHINES.

202. Effects of Centrifugal Force on Bearings—203. Permanent Axis—All Axes of Rotation in Machines ought to be Permanent Axes—204. The Balancing of a Machine—205. Example—Locomotive Engine—Considerations in Designing—206, 207. Balance-Weights on Wheels of Locomotive—208. Rules for Locomotives 218

CHAPTER XXI.

GLOSSARY.

209. Introductory—210. Vertical Line—211. Level Surface—212. Curvature—213. Mass—214. Velocity—215. Acceleration—216. Momentum—217. Impulse or Blow—218. Resultant and Equilibrant—219. Equilibrium—220. Moment of a Force—Law of Moments—221. Example of Moments—222. Torque—223. Lever—224. Couple—225. Work—226. Example of Work—227. Horse-Power—Indicator Diagram—Exercise—228, 229. Energy—Kinetic Energy and Potential Energy—Examples—230. Angle—231. Angular Velocity—232. Angular Acceleration—233. Comparison of Linear Motion and Angular Motion—234. Centrifugal Force—235. Centrifugal Force Apparatus—236. Conical Pendulum—237. Friction—Coefficient of Friction—238. Kinetic Friction Apparatus—239. Friction and Abrasion—240. Friction is often very Useful—241. Fluid Friction—242, 243, 244. Apparatus for Measuring Viscosity of Liquids—Experiments—245. Comparison of the Laws of Fluid and Solid Friction 223

APPENDIX—Rules in Mensuration 252

INDEX 257

LIST OF TABLES.

	PAGE
I.—Moments of Inertia and the **M** of Rotating Bodies	47
II.—Modulus of Elasticity of Bulk, K	57
III.—Melting Points, Specific Gravity, Strength, etc. of Materials	68–9
IV.—Diagrams of Bending Moment, with Strength and Stiffness of a Uniform Beam, when Supported or Fixed and Loaded in various Ways	116–17
V.—Factors of Safety for Different Materials and Loading	121
VI.—Strength and Stiffness of Rectangular Beams Supported at the Ends and Loaded in the Middle	121
VII.—Breaking Stress of Iron and Timber Struts	131
VIII.—Values of the Constant B for Different Lengths of Strut and Different Materials	132
IX.—Values of the Constant n for Struts of Different Sections	132
X.—Normal Pressure of Wind on Roofs	154

PRACTICAL MECHANICS.

CHAPTER 1.

INTRODUCTORY.

1. What I expect the reader to know already.—In this book I mean to consider the principles of mechanics as they are applied in many trades. It is important that I should state in the beginning what is the amount of knowledge which I expect my reader to be possessed of beforehand.

In the first place, he must know the meaning of *decimals* in arithmetic. It is a very strange thing that the meaning of a decimal, which might easily be taught to children before they begin addition, is usually regarded as a part of arithmetic which ought to come after vulgar fractions. Secondly, my reader must own a box of *drawing* instruments; he must be able to set off at once any angle when it is stated in degrees; he must be able to draw a triangle to any scale when one side and two angles, or two sides and one angle, or when three sides are given to him: in fact, he must know as much of the use of drawing instruments and of the use of a scale as a good teacher can give him in one lesson. Thirdly, he must be aware of the fact that a letter of the alphabet or any other *symbol* may be used to represent a physical magnitude. Probably there is nothing more annoying to the person who attempts to give lessons in applied science than the fact that few workmen know the meaning of the simple symbols $+ - \times \div \sqrt{}$. Even

B

when they are aware of the meaning of these symbols they will often get frightened at a mathematical expression, although, if they spent one hour in learning the meanings of such expressions they would never feel afraid of them again. A mathematical expression is simply a very concise way of writing a rule. There are many books of rules, such as Molesworth's, and any workman who knows decimals in arithmetic might be able to use every rule in Molesworth with an hour's study; but he never attempts to learn the key to these secrets, and when he goes to a teacher, it is usually for a long dry course in algebra which he does not really need. Please remember that very few men who use a book of logarithms know how a logarithm is calculated; and just as a man may use a watch or a slide rule, or any other calculating machine, who does not know how to make one, so you may be able to calculate from a mathematical formula, although you do not know how it has been arrived at. I expect that the reason why men have so little practical knowledge of such matters, lies in the fact that their teachers know only one way of studying—the way in which they themselves have been taught—the way of the Universities, which has unfortunately become crystallised in England, and is now in use in Science Classes. It is forgotten that practical men have an experience and an amount of common sense which children have not, and there is a way of giving new ideas to men which we cannot employ with children. This brings me to my fourth requirement, namely, that my reader must know the elementary principles of *mechanics*. If his acquaintance with mechanics is merely derived from books or lectures, he has not the knowledge of which I speak. He cannot know the parallelogram of forces till he has proved the truth of the law half a dozen times experimentally with his own hands. I have met with men who, when given two sides of a triangle in inches, and the angle between them in degrees, could calculate readily the length of the third side and the sizes of the other angles of the triangle, and yet who had never tried with their instruments and drawing paper whether their calculations were correct. Such is not the sort of knowledge which I want my readers to have. I want them to think of things as being measurable, of laws as statements which ought to be submitted to the test of their own experiments, however rough, before they are accepted as true.

2. Equilibrium of Forces.—Take as an example the law called "the triangle of forces"—if three forces act on a small body, and just keep it at rest; then if we draw on a sheet of paper three straight lines parallel to the directions of the three forces, and let them form a triangle, in such a way that arrowheads representing the directions of the forces go round the triangle in the same way, it will be found that the lengths of the sides of the triangle are proportional to the amounts of the forces. Now the statement of this law has very little meaning to the student until he gets three strings and three weights, as in Fig. 1, and by means of pulleys allows the strings to pull at a small body, P, all at the same time. But when, after all sorts of trials of different weights and different positions of the small body, he finds that even his rough tests of the law prove to be satisfactory every time, he has obtained an exact notion of what the law means, which he can never lose again. If, instead of three forces, he lets four or more forces act upon the small body, and, when they balance one another, he draws straight

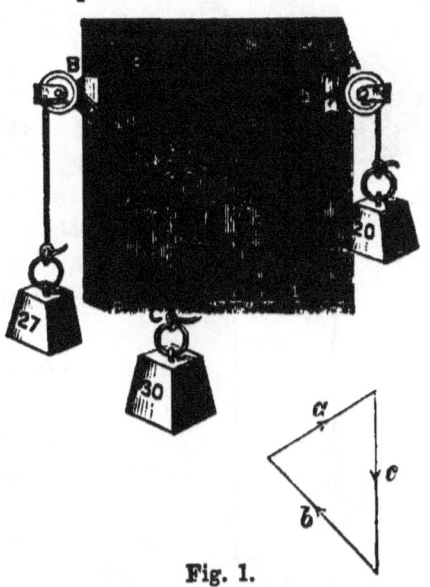

Fig. 1.

lines parallel to the directions of the forces and representing their amounts, according to any scale he pleases, taking care that, in whatever order he draws them, the arrowheads run all in the same *sense*, he will draw a polygon; he will find the polygon to be closed and complete when he finishes, and he will prove roughly the law of "the polygon of forces." He will at once see that the triangle of forces is merely a simple case of the more general polygon of forces.

3. Our one Theory insufficient.—Now, no reasoning man can make these trials without finding that there is a great deal to be observed beyond what his teacher or his book has taught him. A force has been represented by the pull in a string passing over a little pulley with a weight at its end. He finds that as his pulley works more easily, and as its pivots are better oiled, his proof of the law is better and better: in fact, he finds that the pull in a string is not exactly the same on the two sides of a pulley. If he takes one pulley and one string, and two weights, called A and B, Fig. 2, at its ends, he will find that there is equilibrium even when the two weights are not exactly equal. If A is slightly greater than B, and he increases the weight of A till it is just able to overcome B, then the difference between the weights represents what may be called the **friction** of the pulley. If now he increases the weights which he uses, he will find that the friction is proportionately increased, and he will get to understand that this is a general law in machinery: "friction is proportional to load." Again, he sees that this friction, which is a resistance experienced in the *rubbing* together of any two surfaces, is a force which always opposes motion, always acts against the stronger

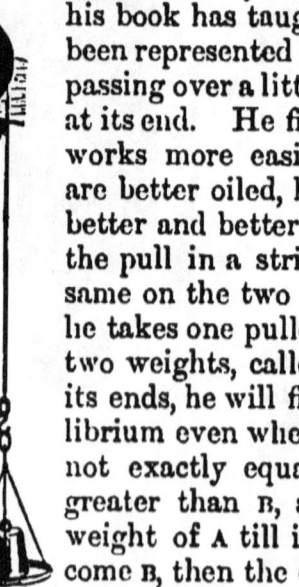

Fig. 2.

influence. Suppose, for example, that he found that a weight of 5·1 ounces was just able to overcome a weight of 5 ounces; he will find that a weight of about 4·9 ounces will just be overcome by a weight of 5 ounces, and that there is equilibrium with 5 ounces and any weight varying from 5·1 to 4·9. **Friction is then a passive force, which always helps the weaker to produce a balance.**

CHAPTER II.

FRICTION IN MACHINES.

4. Law of Work.—Take any machine, from a simple pulley to the most complicated mechanism. Let a weight, A, hung from a cord round a grooved pulley or axle in one part of the mechanism, balance another weight, B, hung from a cord round another axle or pulley somewhere else. In Fig. 3 we have imagined that the mechanism is enclosed in a box, and only the two axles in question make their appearance. Now move the mechanism so that A falls and B rises, and observe their motions. Suppose that when A falls 1 foot B rises 20 feet, then if there were no friction in the machine a weight at A is exactly balanced by one-twentieth of this weight at B. This is the law which you will find proved in books on mechanics. The reason why it is true is this. The work or mechanical energy given out by a body in falling is measured by the weight of the body multiplied into the distance through

Fig. 3.

which it falls. It is in this way that we get the energy derivable from the fall of a certain quantity of water down a waterfall, and it is in this way that we find out whether a certain waterfall gives out enough power to drive a mill (see Chapter VII.). Similarly the energy given to a body when we raise it, is measured by the weight of the body multiplied by the vertical height through which it is raised. Now every experiment we can make shows that energy is indestructible, and consequently, if I give energy to a machine, and find that none remains in it, it must all have been given out by the machine. Therefore the energy given out by A in falling slowly must be equal to the energy received by B in rising, and as A falls 1 foot when B rises 20 feet, the weight of A must be twenty times the weight of B. If, then, there were no friction in the machine, and if a weight of 20 lbs. were hung at A and a weight of 1 lb. at B, we should find that if we start A downwards or upwards there will be a steady motion produced. Any excess at A will cause it to overcome B, the weights moving more and more quickly as the motion continues.

Now, in our machine, Fig. 3, we can always find by trial what is the *velocity ratio;* that is, the speed of B as compared with the speed of A. This cannot alter. But when we try to balance a weight at B by a weight at A, we find that the above relation is quite untrue. Hang a weight of 1 lb. at B, hang a weight of 20 lbs. at A, there is certainly a balance, but when we have somewhat less or more than 20 lbs. at A there still is balance. The reason for this is, that there is friction in the mechanism, and this friction always tends to resist motion, always acts against the stronger influence.

5. **Effect of Friction.**—We shall now proceed to find out in what way friction modifies the law given in the books which I have just spoken about. You must make actual experiments with some machine, if you are to get any good from your reading. Hang on a weight, B, and find the weight, A, which will just cause a slow, steady

motion. Do this when a number of different weights are placed at B. Now, suppose you have measured the velocity ratio, that is, suppose you find that B rises *four* times more rapidly than A falls. Then, according to the books, there would be an exact balance if A were four times the weight of B. On actual trial, however, I find in a special case the following table of values:—

	A overcomes B when				
A is	23·4	ounces and B is		5	ounces
,,	44·7	,,	,,	10	,,
,,	65·4	,,	,,	15	,,
,,	86·8	,,	,,	20	,,
,,	107·5	,,	,,	25	,,
,,	128·8	,,	,,	30	,,
,,	149·6	,,	,,	35	,,
,,	171·0	,,	,,	40	,,

But if there had been no friction-in the first experiment, A would have been 20 ounces instead of 23·4, hence the friction is represented by this 3·4 ounces. For every experiment let this be done, subtract four times B from A and call this difference the friction. Now how shall we compare this friction with the corresponding load?

6. The use of Squared Paper.—And here we come to a matter of the greatest importance to the practical man, which the old-fashioned books on mechanics and old-fashioned teachers of Science Classes seem not to know anything about. How do we practically compare two things whose values depend on one another? How do we find out the law of their dependence? It is a strange fact that there should be a class in the community who have a little difficulty in manipulating decimals in arithmetic, but it is almost a stranger evidence of neglected education that so many people should be ignorant of the great uses to which a sheet of squared paper may be put.

A sheet of squared paper can be bought very cheaply. It has a great number of horizontal lines at equal distances apart, and these are crossed by a great number of vertical lines of the same kind, so that the sheet is covered with little squares. This sheet will enable me first of all to correct for errors of observation in the above

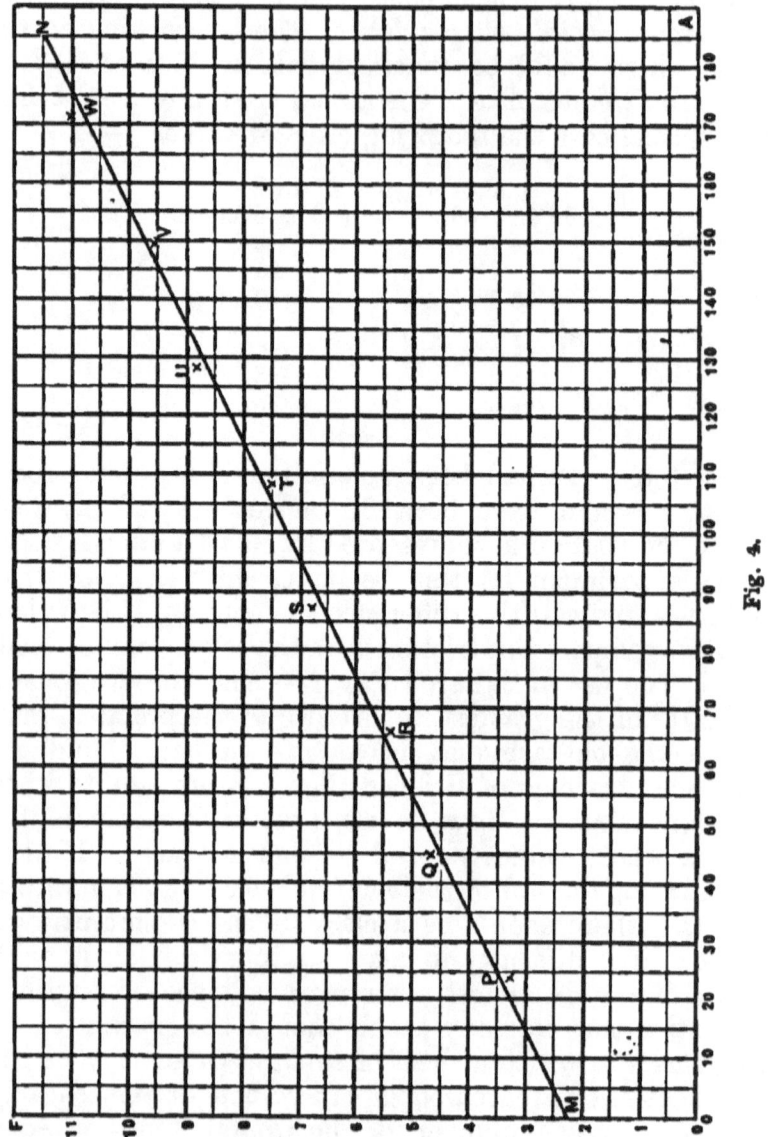

Fig. 4.

series of experiments; and, secondly, to discover the law which I am in search of. A miniature drawing is shown in Fig. 4, many lines being left out because of the difficulties of wood-cutting. At the bottom left hand corner I place the figure 0, and I write 10, 20, &c., to indicate the number of squares along the line, 0 A. Instead of 10, 20, &c., I might write 1, 2, &c., or 100, 200, &c., according to the scale I am going to use. Indeed, on account of the friction being so much less than the weight A with which it is to be compared, I number the squares along the vertical line 0 F by 1, 2, &c. instead of 10, 20, &c. **We can employ any scale we please** in representing either of the things to be compared, and it is usual to multiply all the numbers of one kind by some number, so as to represent all our experiments on one sheet of paper, and on as much of this sheet as possible. Having subtracted four times B from A, I find the following numbers:—

A.	Friction.	A.	Friction.
23·4	3·4	107·5	7·5
44·7	4·7	128·8	8·8
65·4	5·4	149·6	9·6
86·8	6·8	171·0	11·0

I now find on my sheet of paper the point P, which is 23·4 horizontally and 3·4 vertically, and mark it with a cross in pencil. Q is 44·7 horizontally and 4·7 vertically, and so for the others. The last point, W, is 171 horizontally and 11·0 vertically. We guess at the decimal part of a small square. The point P represents my first experiment, and every other point represents one experiment. Now we are certain that if there is any **simple law** connecting load and friction, the points P, Q, to W, lie in a simple curve or in a straight line. You see that, in this case, no simple curve will suit the points; they would evidently lie in a straight line, only that we made some errors of observation. You must now find what straight line lies most evenly among all the points; this you can do by means of a ruler or a fine stretched string, and the line M N seems to me to answer best. It tells me, for instance, that when A is 44·7 the friction is really 4·5, instead of 4·7. Take any point in the line, its vertical measurement gives me the true friction corresponding to a load represented by its horizontal measurement. Thus, for instance, you see that friction 5·0 corresponds to load 55.

This is a simple way of correcting errors made in experiments, but you cannot hope to understand much about it till you actually make experiments and use the squared paper. You will find the matter all very simple when you try for yourself; my description of it is as complicated as if I were teaching you to walk.

7. If, at any time, you make a number of measurements of **two variable things** which have some relation to one another, plot them on a sheet of squared paper, and correct by using a flexible strip of wood or a ruler, to draw an easy curve or a straight line so that it passes nearly through all the points. If the line is straight, the *law connecting the two things* will prove to be a very simple one. In the present case it means that any increase in the load is accompanied by a proportionate increase in the amount of friction. Thus, when the load is 0, the friction is 2·3; when the load is 100, the friction is 7·2. That is, when the load increases by 100 the friction increases by 4·9, so that the increased friction is always the fraction, ·049, of the increased load. In fact, it is evident that we can calculate the friction at any time from the rule

$$\text{Friction} = 2\cdot 3 + \cdot 049\ A.$$

That is, multiply the load A in ounces by ·049, and add 2·3, the answer is the friction.

8. **Law of Friction.**—Our result is that the total friction is equal to the friction, 2·3, of the machine unloaded, together with a constant fraction, ·049, of the load. Now when a similar series of experiments is tried on any machine, be it a watch or clock, or be it a great steam-engine, we always find a similar simple law.

If you clean all the bearings or pivots, or if you use a different kind of lubricator, you will get other values for the two numbers in the above rule, but the law will remain of the same simple kind.

9. Force of Friction.—We have in all this used the term "friction," or the term "effect of friction," to mean the difference between the weight which would balance another through the mechanism if there were no resistance to the rubbing of surfaces, and the weight which will just overcome the other when there is such resistance to rubbing of surfaces. At any rubbing surface there is a force of friction which is proportional to the pressure between the surfaces. At the rubbing surface itself we can speak of this *force of friction*, but when we speak generally of the whole friction of a machine, we are speaking, not of the force of friction at any one surface, for this is different probably from the force of friction at any other rubbing surface, but we speak of an effect which is produced, somehow, by the friction everywhere in the machine, wherever there is a pivot, and wherever the tooth of one wheel rubs on another.

10. Loss of Energy due to Friction.—Thus, in the simple case with which we began (Art. 3), the difference of pull in a cord on the two sides of a pulley was what we called the friction of the arrangement, whereas, really, the friction takes place at every point of the pivot where rubbing occurs, and the force at one point may not be the same as the force at another point. Again, any force acting in the cord has a greater leverage about the axis than any of the forces of friction has. The real connection between the two things is, then, this: what we have generally called "the effect of friction," or "the friction of the arrangement," multiplied by the velocity of the cord on which it is measured, is equal to the sum of all such products as the friction at any point in a rubbing surface multiplied by the velocity of rubbing. In fact, if the weight A in falling cause the weight B to rise, the work done by A is greater than the work done on B by an amount which is called the work lost in friction, and this is the work done against the forces of friction at all the rubbing surfaces.

If we know the force of friction at any place, *in pounds*, and the distance, *in feet*, through which this force is overcome, that is, the distance through which rubbing has occurred, the product of force by distance measures the work or energy spent in overcoming friction, in what is called *foot-pounds*. This energy is all wasted, or, rather, it is all changed into heat and does not come out of the machine as mechanical work, the shape in which it was when we put it into the machine. And inasmuch as no machine can be constructed which will move without friction, we never get out of a machine as much mechanical work as we put into it.

11. Friction at Bearings of Shafts.—At almost every rubbing surface which you can consider, the force of friction is different at every point of the surface, and it is generally acting in different directions at different points. Consider, for example, a horizontal shaft and its bearing (Fig. 5). The force of friction at C, per square inch of area of rubbing surface, is probably not the same as at A. A very little difference in the size of the shaft or its bearing will cause a considerable difference in the pressure per square inch at C or at A.

Fig. 5.

Now the force of friction at C, multiplied by the velocity of rubbing, gives the work or energy lost per second in friction at C; and this, added to the energy lost at every other place where rubbing occurs, gives the total loss of energy per second at all the points. It is not, then, a simple matter to investigate the force of friction at every point of such a bearing; and the rigidity of the metal, and a number of other important matters, must be taken into account in investigating the force of friction everywhere when the shaft is transmitting different amounts of power. As we have already seen, however, experiment shows that the energy lost in friction for a certain amount of motion increases proportionately with the energy actually transmitted by the shaft. Keeping in mind, then, the general law—"the force of friction is proportional to load"—it is easy to see how to reduce the frictional loss in any machine. For instance, when a wheel is transmitting power, the load on the rubbing surfaces of its bearings or pivots depends on the power transmitted. Now, the actual force of friction at the rubbing surface is about the same, whatever be the size of the bearing; but the distance through which rubbing occurs when the wheel makes one revolution is less as we have a less diameter of bearing; in fact, the force of friction, multiplied by the circumference of a cylindric bearing, is the energy in foot-pounds lost in one revolution. Our rule is, then, to make this diameter as small as possible, consistently with sufficient strength. The wheel of a carriage is made large, and the axle, where rubbing occurs, is made as small as possible, because in this way the carriage moves over a great distance for a small amount of rubbing. There is another reason, however, for the use of large wheels in carriages on common roads—namely, their being better able to get over obstacles, such as stones. In some machines, where it is important that there should be very little friction at the bearings of axles, the axles are made to lie at each end, in the angle formed by two wheels with plain rims. The main axle rolls on these wheels, and it is only at the axles of the wheels that there is rubbing. This rubbing is a very slow motion, and as the *force* of friction is but little increased in consequence of the weights of the friction wheels, the energy lost in friction may be made very small in this way.

12. If you compare the **parallel motion** which is still used in some large steam-engines to cause the piston-rod to move in a straight line, with the slide which is now so common, you will see that there is very much less loss of energy by friction when the parallel motion is em-

ployed because, whereas in the slide the rubbing motion is as much as the motion of the piston, in the parallel motion rubbing only occurs at the pivots of the arrangement. Unfortunately, this arrangement does not allow the piston-rod end to move *exactly* in a straight line, and produces some friction between the piston and its cylinder, and between the piston-rod and stuffing-box; and it is also much more costly and less compact than slides. Hence slides are coming into general use.

13. **In quick-moving shafts,** it is usual to make the journals or bearings longer in proportion to their diameter than in slow-moving shafts. This is rendered necessary by the fact that, as all the energy wasted through friction is converted into heat, when there is more power wasted in friction we ought to let the heat get away more rapidly. Giving greater rubbing surface has this effect. Also, as the materials are more abraded when the velocity is too great, lengthening the journal diminishes the pressure at every place, tending thus to counteract the effect of increased velocity.

14. **Mechanical Advantage.**—In books on mechanics you will usually find that, when machines are described, they are only considered in relation to their *Mechanical Advantage*. That is, suppose a small weight, P, usually called the *power*, is able by means of the mechanism to cause a larger *weight*, W, to rise, the ratio of W to P is called the mechanical advantage. Now, in nearly all cases you will find that, when there is a mathematical investigation of a machine, the assumption is made that there is no friction. I have already shown you that the problem of taking friction into account is a very difficult one. But, as we have seen, a practical man can experiment on the effect of friction, and obtain results which the mathematician never attempts to deduce; and, happily for us, these results are generally very simple. Let the reader make a few experiments himself, or let him by means of squared paper find the relation between P and Q from the following results, taken from a crane

whose gearing was well oiled, and whose handle was replaced by a grooved wheel, round which was a cord supporting P:—

W. Weight just Overcome.	P. Power just able to Overcome Weight.
100 lbs.	8·5 lbs.
200 ,,	12·8 ,,
300 ,,	17·0 ,,
400 ,,	21·4 ,,
500 ,,	25·6 ,,
600 ,,	29·9 ,,
700 ,,	34·2 ,,
800 ,,	38·5 ,,

The weight capable of being lifted slowly by the crane we call W. We found that P fell forty times as rapidly as W rose, and you may have imagined that the mechanical advantage was forty, or that a weight, P, could lift a weight, W, forty times as great as itself. This would be true if there were no friction; but we see that in practice it is not the case. Plot the above values of P and W carefully on squared paper, and you will find that, if the weight W is increased 1 lb., P must be increased ·0429 lb.; and also that when W is 0, a power, P, of 4·21 lbs. is needed to cause a slow motion of the crane; so that the law is

$$P = 4·21 + ·0429 \, W.$$

Namely, multiply the weight W in pounds by the fraction ·0429, and add 4·21: the answer is the power required to lift W. When you have worked out this rule, employ it in finding how much power, P, is required to lift a ton with such a crane. Answer, 100·3 lbs.

15. Rate of doing Work.—I have been using the word *power* in a very wrong sense in Art. 14, because you will find it used in this way in books on mechanics. I have used it as a mere name for the weight P, which causes the weight W to rise. But the word *power* will be used by us in future in a very different sense. If a weight of 1,000 lbs. fall 100 feet in two minutes, it gives out 1,000 × 100, or 100,000 foot-pounds of work in two minutes, or 50,000 foot-pounds of work in one minute. Now, 33,000 foot-pounds of work done in one minute is called *a horse-power*, and hence our falling weight gives out $50,000 \div 33,000$, or 1·5 horse-power. Ten horse-power means ten times 33,000 foot-pounds of work done in one minute. The idea, then, of *power* is an idea of work done in a certain time.

16. Economical Efficiency.—Take any pair of numbers from the above table, say $P = 8\cdot5$ lbs., when $W = 100$ lbs. Let us suppose that P is moving at the rate of forty feet per second, then we know that W is rising at the rate of one foot per second. P is giving out the power $8\cdot5 \times 40$, or 340 foot-pounds per second; W is receiving 100 foot-pounds per second. The ratio of the power usefully employed to the power given to the machine is called the *efficiency* of the machine, so that our crane has an efficiency $100 \div 340$, or ·294. Sometimes the efficiency is put in the form of a fraction; sometimes we say that it is 29·4 per cent., meaning that it employs usefully 29·4 per cent. of the work given to it.

Now take another pair of numbers, say $P = 38\cdot5$, $W = 800$, and let P fall forty feet in one second, as before. We now get as our answer ·519—that is, more than half, or 51·9 per cent., of the power given to the crane is usefully employed. We see, then, that as the power given to the crane is greater for a given speed, the efficiency is also greater. This arises from the fact that the friction of the unloaded crane is always entering into the calculation; and if we take the case where no weight, W, is being lifted, and P must be 4·21 lbs., we shall find an efficiency, 0, because work is being given to the crane, and none is coming out usefully. You will always find that the power usefully given out is a certain fixed fraction of the total power given to the machine, minus the power required to drive the crane at the given speed when it is unloaded. Choose some speed, say that P falls forty feet per second; find the total power or 40 P; find

the usefully employed power $1 \times w$ for every case of the above table. Plot your answers on squared paper, and you will find this law :—
Total power = useful power × 1·716 + { Power required to drive crane at same speed when unloaded.

CHAPTER III.

MACHINES—SPECIAL CASES.

17. Blocks and Tackle.—It is very good to have a general law telling us about machines in which there is no friction. That law you now know. *The mechanical work given to a machine is equal to the work given out by it,* unless it is stored up in the machine itself by the coiling of a spring or in some other way. But, besides knowing the law itself, it is well to know what it leads to in certain special cases. Take, for instance, a pulley-block, Fig. 6. It is evident here that if we have three pulleys in the block B, if the cord P is pulled *six* inches, W will only rise *one* inch, and therefore P will balance six times its weight at W if there is no friction. The mechanical advantage is therefore six.

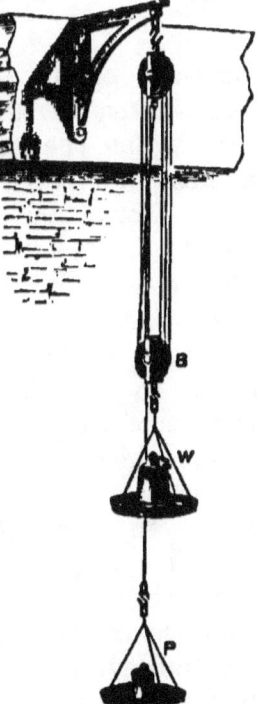

Fig. 6.

18. Inclined Plane.—Again, take the inclined plane, Fig. 7; W is a weight which will roll down the plane without friction, let us suppose; P is the pull in a cord which just prevents W from falling. The cord is parallel to the plane. Evidently when W rises from level A to level B the cord is pulled the distance A B; that is, W multiplied by the height of the plane is equal to P multiplied by

C

the length of the plane. Thus, if W is 1,000 lbs., and the length of the plane 10 feet for a rise of 2 feet, then ten times P is equal to 2,000, or P is 200 lbs.

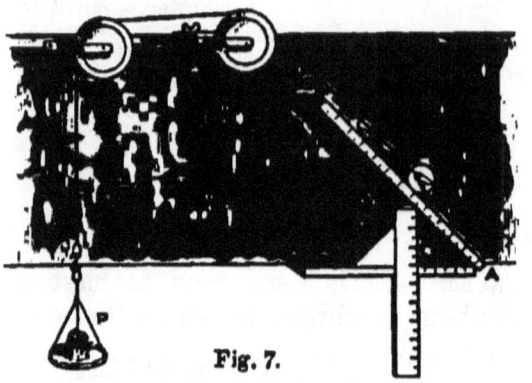

Fig. 7.

19. **The Screw.** —Again, suppose there is no friction in the screw A B, Fig. 8, if it rise it lifts a weight say of 3,000 lbs. Now, if the screw make one turn it rises by a distance equal to its *pitch*, that is, *the distance between two threads.* Say that the pitch is ·02 foot, then when the screw makes one turn it does work on the weight 3,000 × ·02, or 60 foot-pounds. But to do this, P must fall through a distance equal to the circumference of the pulley A, about which I suppose the cord to be wound. Suppose the circumference of the pulley to be 6 feet, then P multiplied by 6 must be 60, or P is 10 lbs. The rule, then, for a screw is this—*power multiplied by circumference of the pulley equals weight multiplied by pitch of screw.* It is not usual to have a pulley and a cord working

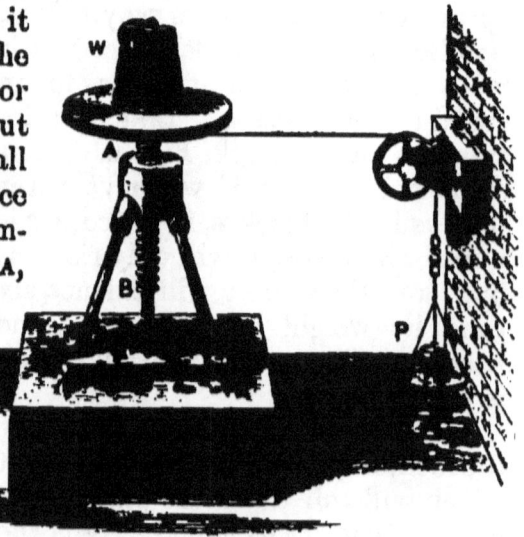

Fig. 8.

Chap. III.] DIFFERENTIAL PULLEY-BLOCK.

a screw; it is more usual to have a handle, and to push or pull at right angles to the handle. Instead of the circumference of the pulley, we should take, then, the circumference of the circle described by the point where the power is applied to the handle.

Exercise.—A steam-engine gives to a propeller shaft in one revolution 60,000 foot-pounds of work; the pitch of the screw is 12 feet. What is the resistance to the motion of the vessel? Answer: The resistance in pounds multiplied by 12 gives the work done in overcoming this resistance, and this work (leaving friction out of account) must be equal to 60,000 foot-pounds, hence the resistance to the motion of the vessel is 5,000 lbs. (We have here assumed that there is no *slip* in the screw.)

20. A differential pulley-block is shown in Fig. 9. When the chain A is pulled, it turns the two pulleys, or rather one pulley with two grooves, B and C. Now C is a little smaller than B, so that, although at D the chain is lifted, it is lowered at E. If the circumference of B is 2 feet, and that of C is 1·99 foot, then, when A is pulled 2 feet, D is lifted 2 feet, but E is lowered 1·99 foot, hence the pulley F, although it will turn a considerable distance, will only rise 0·01 foot, carrying the weight W with it. If W is 2,000 lbs., then 2,000 × ·01, or 20, must be equal to P, the pull in A, multiplied by 2, hence P is 10 lbs., or a power of 10 lbs. is able to lift a weight of 2,000 lbs. The general rule, then, for the differential pulley-block is, *power multiplied by circumference of larger groove* B *is equal to weight multiplied by difference between the circumference of the two grooves* B *and* C.

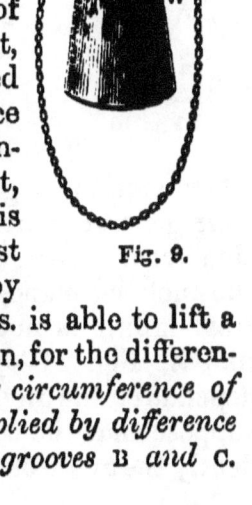

Fig. 9.

You will find that this rule comes to the same thing—*power multiplied by diameter of* B *is equal to weight multiplied by the difference between the diameters of* B *and* C. The grooves are furnished with ridges to catch the links of the chain, so that there shall be no slipping.

21. Wheel and Axle.—If A and B, Fig. 10, are two pulleys or drums on the same axis and having cords round them, a small weight, P, hung from A, will balance a larger weight, W, hung from B. For, suppose that one complete turn is given to the axis, P falls a distance equal to the circumference of A, whilst W is rising a distance equal to the circumference of B. Hence

P × circumference of A = W × circumference of B,

or, what really comes to the same thing,

P × diameter of A = W × diameter of B,

or P × radius of A = W × radius of B.

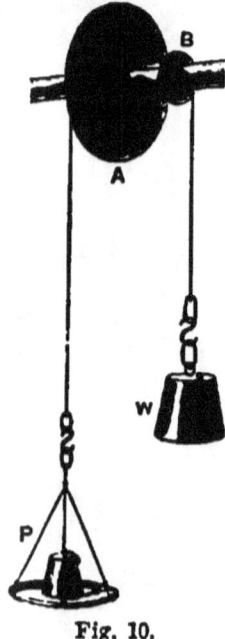

Fig. 10.

22. Equilibrium in one Position.—In all the machines which we have hitherto considered, we could give motion without altering the balance of P and W, but there are many machines in which the mechanical advantage alters when motion is given. In such cases you will employ your general principle, but you must make your calculation from a very small motion indeed. For instance, in the inclined plane, if the cord which prevents the weight from falling is not parallel to the plane—say that it is like M, Fig. 11—you will find that the necessary pull depends on the angle the cord makes with the plane. Now, suppose that the cord pulls the carriage from *b* to *c*, evidently the angle of the cord alters. The question is, what is

EQUILIBRIUM IN ONE POSITION.

P, that it may support W in the position shown in the figure? We know that it will be different after a little motion, but what is it now? Imagine such a very small motion from b to c to occur that the angle of the cord does

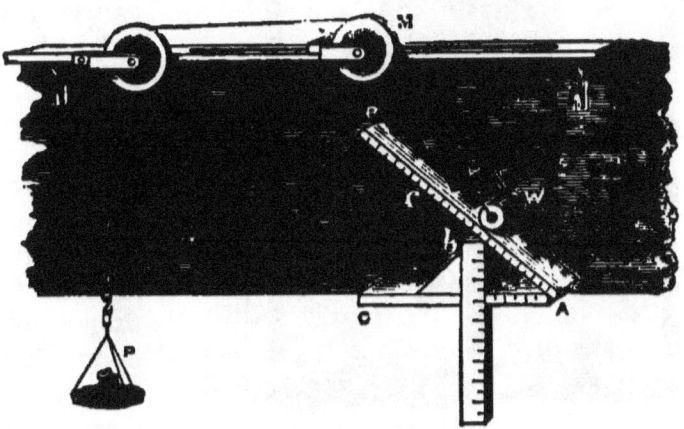

Fig. 11.

not alter perceptibly, and now make a magnified drawing, Fig. 12. P has not fallen as much as the distance $b\ c$, it has only fallen the distance $b\ a$ ($c\ a$ is perpendicular to $b\ a$). In the meantime the weight has been lifted the distance $k\ c$. Hence,

W × $k\ c$ ought to be equal to P × $b\ a$.

Thus, if you measure $k\ c$ and $b\ a$ on your magnified drawing to any scale you will find the relation between P and W. Another way of finding the same relationship is this. We know that the weight of W acting downwards, the pull in the cord, and a force acting at right angles to the plane, are the three forces which keep W where it is. Draw a triangle whose three sides are parallel to the directions of these three forces, Fig. 13, with arrow-heads going round in the same way; then x and y are in the proportion of W and P. Here we have used the principle called "the triangle of forces" to find P.

23. Body turning about an Axis.—In Fig. 14 we have a body which can move about an axis. It is acted on by a number of cords exerting forces which just balance one another. Now, if you make this experiment you will find that you must keep your finger on the body, because it is in such a state that a very small motion either way causes the forces to no longer balance. Suppose, however, you were to let the cord A be wound on a pulley whose radius is equal to the distance A O; the cord B on a pulley whose radius is equal to B O, and so on, you would have the arrangement shown in Fig. 15, which differs from Fig. 14 in that a small motion has no effect on the balance. Now what is the condition of balance in this case? Suppose one complete turn given to the axis,

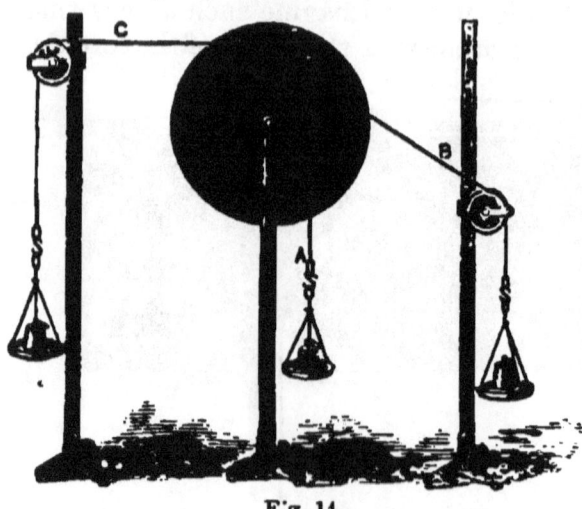

Fig. 14.

Fig. 15.

every cord shortens or lengthens by a distance equal to the circumference of the pulley on which it is wound. Let A and B lengthen, and let C shorten, then we know that the work done by A and B must be equal to the work done against C. Hence,

> Pull in A × circumference of A's pulley, together with pull in B × circumference of B's pulley, must be equal to pull in C × circumference of C's pulley.

We might, however, use the diameters or radii of the pulleys, and so we see that in Fig. 14 there is balance if

> Pull in A × AO together with pull in B × BO, equals pull in C × CO.

The pull in A × AO is really the tendency of A to turn the body about the axis, and in books on mechanics it is called the **moment of the force** in A about the axis O. The law is then, if a number of forces try to turn a body and are just able to balance one another, the sum of the moments of the forces tending to turn the body *against* the hands of a watch must be equal to the sum of the moments of the forces tending to turn the body *with* the hands of a watch.

24. The Lever.—Thus, for example, a lever is a body such as I have spoken about, capable of turning about an axis. You will find that our general rule of work, and this rule of moments, will give the same result. *If two forces act on a lever, they will balance when their moments about the axis are equal;* that is, when P, multiplied by the shortest distance from the fulcrum or axis to the line in which P acts, is equal to W multiplied by the distance of the fulcrum from the line in which W acts.

If a number of forces balance when acting on a lever, the sum of the moments tending to turn the lever against the hands of a watch must be equal to the sum of the moments tending to turn the lever with the hands of a watch.

It must be remembered that, if the body acted upon

has its centre of gravity somewhere else than in its axis, then we must consider that the weight of the body is a force acting through its centre of gravity.

Exercise.—The safety valve, Fig. 16, must open when the pressure on the valve is just 100 lbs. per square inch. The mean area of the valve A, on which we assume that the pressure acts, is 3 square inches; CD is 2 inches, E is 50 lbs., the weight of the lever is 6 lbs., and its centre of gravity is 6 inches from D—where must E be

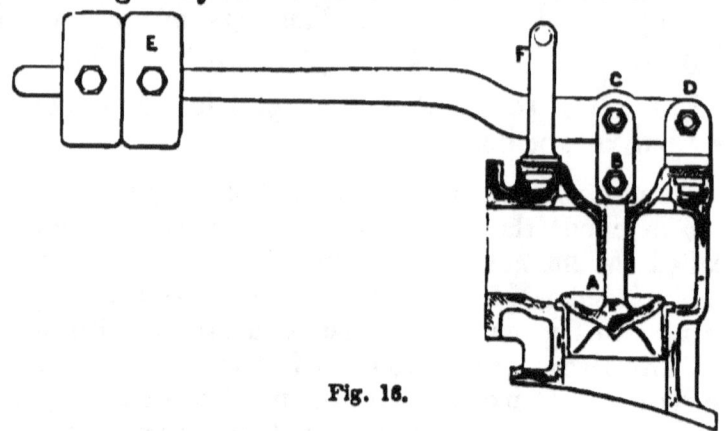

Fig. 16.

placed? Here the upward force is 100 × 3, or 300 lbs., and its moment about D is 300 × 2, or 600. The moment of the weight of the lever is 6 × 6, or 36. The moment of the weight E is 50 × the required distance from D. Hence, 600 — 36, or 564 divided by 50, is the answer; 11·28 inches from D.

Exercise.—A weighbridge consists of three levers whose mechanical advantages help each other; I mean, the short arm of each supports the long arm of the next. Suppose that the weights of all parts are arranged so as just to be balanced when no weight is on the bridge, and that the mechanical advantages of the three levers are 8, 10 and 12, what weight will be balanced by a power of 15 lbs.? Answer, 14,400 lbs. Suppose that it is the first of these levers that is alterable (that is, the power is a sliding weight), what is its mechanical advantage altered to when the load is 16,000 lbs.? Answer, It was 8, it now becomes increased in the proportion of 16,000 to 14,400, so that it becomes 8·8889 feet.

25. Hydraulic Press.—A hydraulic press is a machine which enables great weights to be lifted, or

Fig. 17.

great pressures exerted, but in which, instead of levers and wheels, we use water to transmit the energy. In Fig. 17 the labourer exerts a force of, say 20 lbs. at P. If P o is 15 times Q o, then Q and the plunger Q s move at one-fifteenth of the speed of P. Now, let us suppose that the labourer has been working for a few minutes, the water filling the whole tube and cylinder space from the ram M to s may be regarded as incompressible, and the cylinder BB is unyielding; and if we force the plunger into this space, the ram must rise if there is no leakage. The cubic contents of the water displaced by the plunger must somehow or other be provided for, and it is the motion of the ram which provides for it. If the ram has an area of cross section of 200 square inches, and the plunger has an area of cross section of only one square inch, then the plunger must move 200 times as rapidly as the ram; hence, the hand of the labourer must move 15×200, or 3,000 times as rapidly as the ram. Hence, if there were no friction, the ram would lift a weight of $20 \times 3,000$, or 60,000 lbs. *The mechanical advantage of the hydraulic press is then found by multiplying the area of cross section of the ram by the mechanical advantage of the lever, and dividing by the area of cross section of the plunger.*

CHAPTER IV.

MACHINERY IN GENERAL.

26. Mechanism.—When the power of a steam engine is distributed through a factory, the distribution is performed by means of shafts, spur and bevil wheels, belts and pulleys, and other kinds of gearing. As I am writing for men who have observed such transmission of energy, it is no part of my object to describe here what can be seen in any workshop. Perhaps no study is more useless from books alone than the study of mechanism; whereas, it is very useful and easy if you examine the actual thing, or make a skeleton model or a skeleton drawing. What I shall say, then, is to help you in your observation rather than to give you a knowledge of mechanism.

27. Velocity Ratio.—In any machinery the velocity of any point may be calculated when the velocity of any other point is known. The number of revolutions per minute made by a shaft tells us the velocity of any point on any wheel or pulley fixed on the shaft; the circumference of the circle described by such a point, multiplied by the number of revolutions, is evidently the distance moved through by the point in one minute. Now, when one shaft drives another by means of spur or bevil wheels, or by two pulleys and a strap, it is evident that the number of revolutions per minute made by one of the shafts, multiplied by the number of teeth of the wheel, or by the circumference or diameter of the wheel or pulley, is equal to the number of revolutions made by the other shaft, multiplied by the number of teeth, or by circumference or diameter of the other wheel or pulley. This is evidently true, supposing that the strap does not slip on the pulley. Hence the rule—to find the speed of a shaft, driven from another by means of any number of wheels or pulleys, *multiply the speed of the driving shaft by the product of the diameters or numbers of teeth in all the driving wheels or pulleys, and divide by the product of the diameters or numbers of teeth in all the driven wheels or pulleys.* By the diameter of a spur wheel we mean the diameter of its *pitch circle*. Two spur wheels enter some distance into one another, and the circle on one which touches a circle on the other, the diameters of these circles being proportional to the numbers of teeth on the wheels, is called the pitch circle. The circumference of the pitch circle, divided by the number of teeth, gives the *pitch* of the teeth.

28. Shapes of Wheel Teeth.—We know that if two spur wheels gear together, however badly their teeth are formed, so long as a tooth in one drives past the line of centres of a tooth in the other, their average speeds follow the above rule. But if we want the speed at any instant to be the same as at any other instant, it is necessary to form the teeth in a certain way. *The curved sides of teeth ought to be cycloidal curves.* The proof of this is not very difficult, but I shall not give it to you. It is not usual to employ these cycloidal curves, for it is found that certain arcs of circles approximate very closely to the proper curves. The method of drawing rapidly the curved tooth of a wheel you will find taught by every teacher of mechanical drawing, you will find described in a great number of books, and you will see it in use in the workshop.* You must remember that no study of books, and I may also say, no fitter's or turner's work that you may engage in, will make up for want of the experience which you would gain by actually drawing to scale a spur or bevil wheel, a bracket or pedestal with brasses, and a few other contrivances used in machinery. **A worm and worm-wheel**, that is, a screw, every revolution of which causes one tooth of a wheel to be driven forward, is sometimes used when we wish to drive a shaft with a very slow speed. If the worm-wheel has 30 teeth, it evidently makes one-thirtieth of the number of revolutions of the driving shaft.

29. **Skeleton Drawings.**—When we consider the relative motions of, say, a piston and the crank which it drives, we come to something which it is not so easy to state without some little knowledge of mathematics. It is the same with all sorts of combinations of link work, and with cams. Even a good knowledge of mathematics is only sufficient to give one a rough general idea of the relative motion in such cases; and for the study of any special case there is nothing so good as a skeleton drawing or a model. I give one example of the use of skeleton drawings—a **crank and connecting rod**. Let A and B, Fig. 18, be the ends of a connecting rod. As A moves from a_1 to c and back again, B describes the complete circle, $b_1 d b_1$. Set off equal distances to b_1, b_2, b_3, &c., and make $b_2 a_2, b_4 a_4$, &c. equal to the length of the connecting rod. Then the points a_1, a_2, &c. and b_1, b_2 &c. show

* Consult Professor Unwin's *Machine Design* on the teeth of wheels.

CRANK AND CONNECTING ROD.

in a very good way the relative motions of A and B. When you have finished this exercise, work others in which, with the same length of crank, you have longer or shorter connecting rods. You will get some such results as are shown in the upper part of the figure. In every case, if we imagine the crank to revolve uniformly, the motion of A, the end of the connecting rod, is shown; the distance from one point to the next is passed over by A in the same interval of time. **Pure Harmonic Motion** (see Chap. XVII.) is the name given to the motion of the piston rod, when we imagine the connecting rod to be infinitely long; or rather, as we make the connecting rod longer and longer, we get more and more nearly to this sort of motion. You see, then, that by skeleton drawings I mean drawings which show successive positions of the different parts of a mechanism whose motions you want to study. You will find that an **eccentric** and its rod may be regarded as a crank (the length of the crank is the distance between the axis about which the eccentric is revolving and its true centre), and a very long connecting rod (the length of the connecting rod being the length of the eccentric rod measured to the true

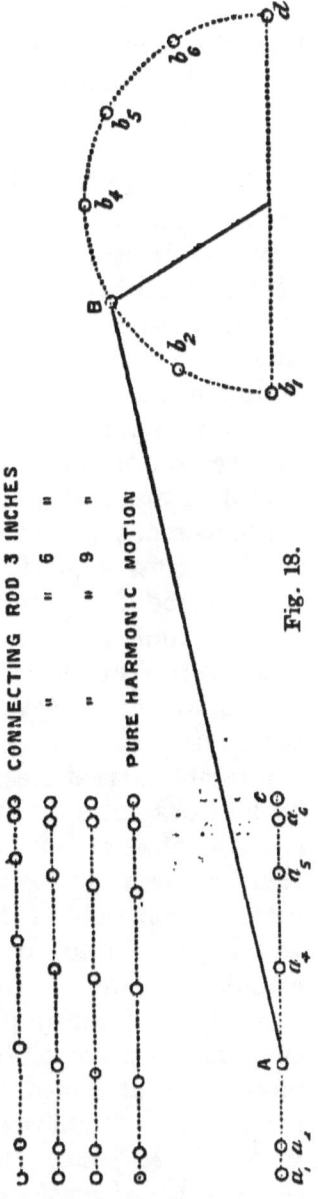

Fig. 18.

centre of the disc). The advantage derivable from skeleton drawings will be more obvious if you consider, in the above case of a crank and connecting rod, that A need not be the cross-head at the end of a piston rod; it may be the end of a lever, and so move in the arc of a circle; it may be a slide moving in a slot of any curved form. One of the most instructive cases of skeleton drawing is a link motion. Taking any good drawing of a link motion to start with, find the relative motions of piston and of slide valve for various positions of the link. In the study of the motion of a slide valve it is much too usual to assume that the piston's motion is what is shown in Fig. 18 as pure harmonic motion. The reason of this lies in the ease with which it can be stated in mathematical language; but it is incorrect, and leads to many errors.

30. **How a Shaft transmits Power.**—I have refused to describe for you what you may see for yourselves at any time in workshops, how spur and bevil wheels and belts transmit power; how there are arrangements for disengaging such gearing, and stepped cones for giving change of speed when belts are used; how shafts are carried near walls or columns; how machine tools work, and a hundred other matters about which a little observation and drawing is of more importance than a large amount of reading. But there are some matters, connected with machinery, of great interest to you which you are not likely to observe unless I direct your attention to them. When a shaft transmits power it is in a state of strain; it is in a twisted condition. The twist is not perceptible to the eye, of course, but methods have been arranged to show it to the eye and measure it; and it is found that *the twist in a shaft is proportional to the horse power transmitted by the shaft divided by the number of revolutions per minute.* Now to explain what I mean by a *twist*. Let a straight line be drawn along the shaft when power is not being transmitted, then, if power be transmitted, the shaft will receive a

twist, and this line will become a spiral line. The inclination, at any point, of the spiral line to its old position, is a measure of the twist.* When, instead of the ordinary coupling, Fig. 19, in which the two halves are connected by means of bolts, we use one, Fig. 20,† in which the two halves are connected by means of spiral springs, these springs get extended when the shaft transmits power. The yielding of the springs cannot be observed unless we make some arrangement like that shown, where a motion of A relatively to C causes the arm E to move and bring the bright bead B towards the axis. If everything is made dead black except the bead it will be seen describing a circle of greater or smaller radius, and a scale with a sliding pointer enables us to measure accurately the distance moved inwards by the bead. *The reading on the scale multiplied by the number of revolutions of the shaft per minute, tells us at once the horse power actually passing through the coupling.*‡

Fig. 19.

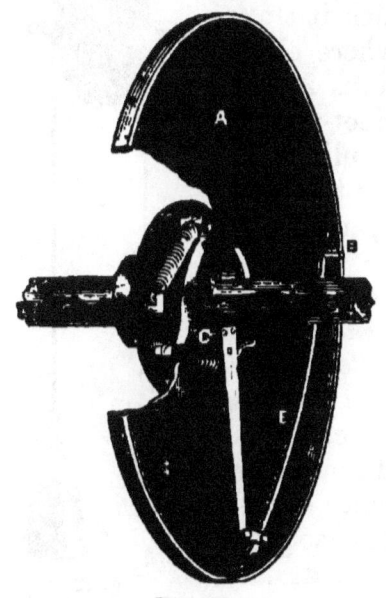

Fig. 20.

31. Belts.—If the pulley A, Fig. 21, is driven from B by means of a belt, you must remember that there is a pull in the part of the belt M,

---

* The best measure of the twist is this angle of the spiral divided by the radius of the shaft, and the quotient is called the *angle of twist*. See Art. 91.

† Ayrton and Perry's Dynamometer Coupling.

‡ The total moment of the forces of the springs in pound-feet,

as well as in the part N. These two pulls are generally pretty great, as you know, but if you could measure them accurately you would find that there is more pull in N, else A would not turn. It is the *difference of these pulls* which concerns us. You may perhaps understand this better from Fig. 22. The pull in A M is the weight of M, say, 20 lbs. The pull in A N is the weight of N, say, 50 lbs. If N falls two feet, M rises two feet, and the work done upon the pulley and which it transmits through the shaft somewhere else is 50 × 2, or 100 foot-pounds, minus 20 × 2, or 40, the difference being 60 foot-pounds. In fact, it is the difference of pull in the two cords, 30 lbs., multiplied by the space passed over by the cord, 2 feet; result, 60 foot-pounds.

Fig. 21.

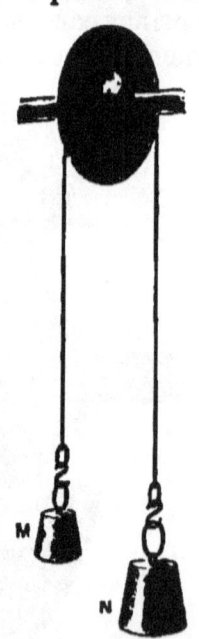

Fig. 22.

The horse power given by a belt to a pulley is then the difference of pull in the belt on the two sides of the pulley, multiplied by the speed of the belt in feet per minute, divided by 33,000. (See Horse Power, in the GLOSSARY.) And now comes the question—if it is the difference of pull that produces turning, why is there so great a pull even in M, Fig. 21, as we usually

or, as it has been called by Professor James Thomson, *the torque*, multiplied by the angular velocity per minute, divided by 33,000, is the horse-power. Suppose that when one of the lengths of shafting is held fast we find the position of the bead when we hang weights on levers or round pulleys or wheels fastened to the other length; a torque of 52·5 pound-feet will cause the bead to move radially inwards by a distance which we call ·01 on our scale; a torque of 105 pound-feet causes the bead to move inwards a distance which we call ·02 on our scale, and so on.

find? Refer again to Fig. 22. If we want the difference between M and N to be 30 lbs., why not make M have no weight at all, and N may then be only 30 lbs? Evidently we should not be able to get friction enough, and the weight N would fall, causing the cord to slide on the pulley; in fact, the friction between the cord and pulley must be more than 30 lbs., else there will be slipping; and to produce this friction it is necessary to have a weight at M as well as at N. If we allowed the cord to lap round more of the pulley the necessary friction might be produced with a less weight at M. To get an idea of the friction between a cord and a pulley, arrange a pulley, or other round object, P, as in Fig. 23. Fix it firmly. Place a weight at M, say 1 lb. Now place weights in the scale pan at N, until the cord just slips slowly. Say we find 3 lbs. to be necessary. The difference between N and M, or 2 lbs., is the friction. Now put twice the former weight at M; you will find that about twice

Fig. 23.

the former N will just cause slipping, so that the friction is doubled. In fact, we have our old law, "friction is proportioned to load." But now let us see **how friction depends on the amount of lapping** of the cord. In your first experiment measure the cord actually in contact with the post P. Suppose it to be 4 inches:

D

now, keeping M, 1 lb., let the cord lap round more of the post P, say 8 inches this time, and find the weight, N, which will just produce a slow sliding. You will find it to be 9 lbs. If the cord touches on 12 inches of the post P, you will find that 27 lbs. at N will be necessary to slowly overcome the friction. It is only by actually trying this experiment for yourself, that you will get a clear idea of how rapidly the friction increases with the amount of lapping. It is on this account that one man can check the motion of the largest vessel by simply coiling a rope a few times round a post.

The apparatus, Fig. 23, is so arranged that any required amount of lapping may be given to the cord round the fixed post P. In an actual experiment, the fixed weight M was 50 grammes. By means of the pulleys the amount of lapping round P was varied, and weights were placed in N, in each case just sufficient to overcome the friction and raise M slowly, as above described. The following are the results of the whole series of experiments:—

Number of times the cord laps round.	The weight required at N to overcome friction and the weight of M.	Logarithms of the ratio of N to M.
$\frac{1}{2}$	80	0·2041
$\frac{3}{4}$	105	0·3222
1	150	0·4771
$1\frac{1}{4}$	200	0·6021
$1\frac{1}{2}$	255	0·7076
$1\frac{3}{4}$	330	0·8195
2	400	0·9031
$2\frac{1}{4}$	500	1·0000
$2\frac{1}{2}$	700	1·1461
$2\frac{3}{4}$	1,000	1·3010
3	1,150	1·3617
$3\frac{1}{4}$	1,500	1·4771

Plotting the first and third columns on squared paper (see Arts. 6 and 7), we find that a straight line passes

nearly through all the points. From this line we deduce
the equation—
$$n = 2 \cdot 2 \log \tfrac{N}{M},$$
where n is the number of times the cord laps round.
From this it is easy to show that the coefficient of
friction, k, between the cord and the post is ·166.*

You must then remember, that the tension in M,
Fig. 21, is necessary to produce as much friction as
will prevent slipping. If ever the excess pull in N is
greater than the friction, there will be slipping. If the
belt slips, there is energy wasted, which you can calculate
if you know the force of friction, and multiply by the distance through which slipping occurs.

32. Transmission and Absorption Dynamometers.
—I have already described to you an instrument which
allows us to measure the horse power transmitted by a
shaft. I am in the habit of employing a somewhat
similar arrangement for measuring the power transmitted
by a belt to any machine. It is shown in Fig. 24, and
is easily understood from the description of Fig. 20. I
can take it near any machine, and drive the machine
through it, using two belts instead of one. G is a loose
pulley. A belt drives H, which drives the plate E
through four spiral springs B. The plate E is keyed to a
shaft carried on the frames C and D, and the pulley F
is keyed on the shaft. A belt from F, therefore, will
drive any machine. When much torque is acting, the
springs B become extended, causing a relative motion of
E and H, and this motion is shown by the bright bead
A, at the end of the lever I A, approaching the axis of
rotation. A fixed scale attached to the frame C allows
the motion of A to be measured.

* The law is this. If k is the coefficient of friction between the
cord or belt and the pulley; if l is the length of the cord or belt
which touches the pulley, say in inches; and r the radius of the
pulley in inches;—then
$$\text{Log } \tfrac{N}{M} = 0 \cdot 4343 \, k \tfrac{l}{r}$$
N and M being the pulls in the belt or cord on the two sides of the
pulley.

36 PRACTICAL MECHANICS. [Chap. IV.

I employ another kind of instrument to measure the horse power given out by any steam-engine, or other

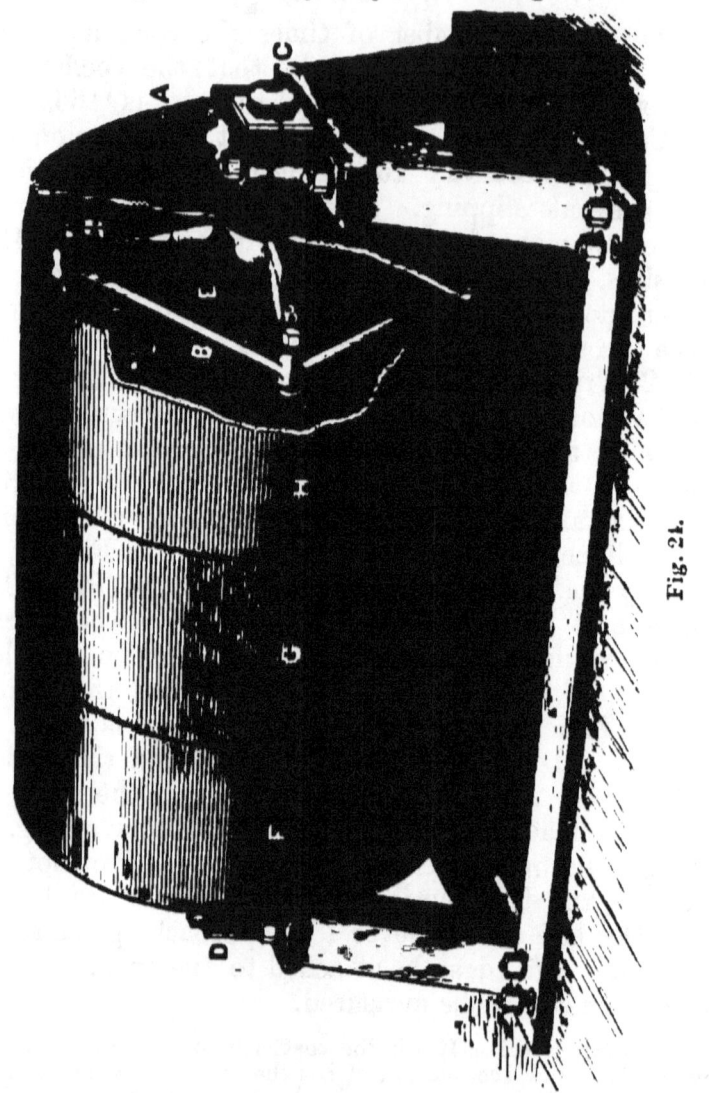

Fig. 24.

motor. The steam-engine drives the pulley A, Fig. 25,* and the pulley B turns along with A. A cord hangs lapping round part of B, and carries at its one end a scale pan, M,

* Charpentier's or Professor James Thomson's Dynamometer.

containing a weight. The other end, N', is pulled by means of a piece of metal fastened to the rim of a loose pulley, C, which has a weight, N, always acting upon it, tending to turn it round. Evidently the cord is pulled with a weight, M, at one end, and a weight, N, at the other. If now there is slipping between the cord and B, the friction is measured by the difference of the weights N and M. If M is 1,000 lbs. and N is 4,000 lbs. the friction is 3,000 lbs. If the pulley has a circumference of 2 feet, and makes 80 turns per minute, the amount of slipping is 80 × 2, or 160 feet per minute, and the work done against friction is 160 × 3000, or 480,000 foot-pounds per minute, that is, 14·545 horse power. In this case *all the power is wasted in friction*, and this is called an Absorption Dynamometer because it measures the power but absorbs it in doing so; whereas the coupling of Fig. 20 and the dynamometer of Fig. 24 are called Transmission Dynamometers, because they *measure the power transmitted through them whilst working any machines*. Any alteration in the torque is shown by a change in the amount of lapping of the cord N' B, and one of the weights must be altered if the speed is to be maintained constant.

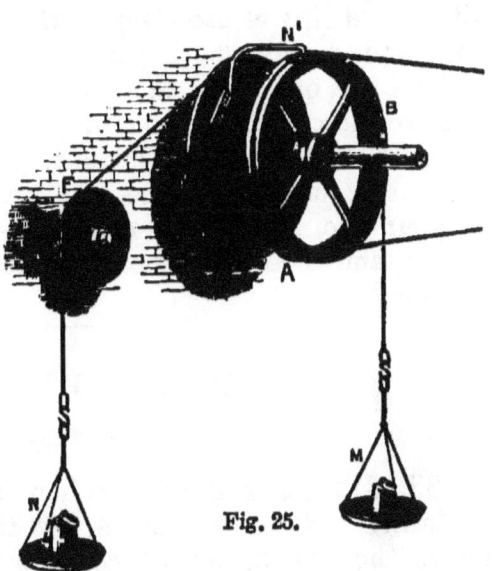

Fig. 25.

You will ask, perhaps, why we do not simply put a rope round B, with spring balances or weights at each end? The answer is, because of slight alterations in speed, little vibrations in the cord and changes in the coefficient of friction; these produce large effects, and you would find that even if you used a dash pot to still the vibrations, the readings on the balances would continually alter; and if you use weights they will jump about in a dangerous manner.* Again, you must take two readings instead of one. In the absorption instrument which I have described to you, if the coefficient of friction diminishes there is an instantaneous alteration in the amount of lapping of the cord on B which is invisible to your eye, but which makes the weights keep quite steady, and their difference is an accurate measure of the friction.

CHAPTER V.
FLY-WHEELS.

33. Kinetic Energy.—When a weight, A, Fig. 3, in falling lifts a weight, B, by the use of a machine inside the box C, let us consider the store of energy at any instant. The store of energy consists in— First. The *potential energy* of A, that is, the weight A in pounds, multiplied by the distance in feet through which it is possible to let it fall. Second. The potential energy of B, which is the weight of B multiplied by the distance through which it is possible to let B fall. Third. The *energy of motion*, or *kinetic energy*, of everything which is moving, namely, A, B, and the parts of the mechanism. We are supposing that there are no other weights which can fall or rise, and that there are no coiled springs or other stores of energy in the mechanism. Now, if A is just heavy enough to maintain a steady motion, the kinetic energy remains the same; so that, whatever energy is given out by A in fall-

* This effect has been observed by several well-known experimenters. However, in recent experiments made (August, 1882) since the above was written, we have found that a very slight guiding of one end of the rope, with the hand or otherwise, so as to keep the whole rope in a plane perpendicular to axis of rotation, is quite sufficient to prevent any jumping, without interfering with the accuracy of the observations. We used a spring balance at one end.

ing is in part being given as potential energy to B, and is in part being wasted in friction. But suppose A to be heavier than this, then there is more potential energy being lost by A than is being stored by B or wasted in friction, and it must be stored up in some other form. The surplus stock shows itself in a quicker motion of everything; it is being stored up as kinetic energy.

34. Energy Indestructible.—We have now to consider an important question. When a certain amount of potential energy (measurable in foot-pounds) disappears, and becomes kinetic energy, how quickly must all the parts of the machinery move to store it all up? This problem is very troublesome, because everything in Fig. 3 is in motion in a different way; some parts of the mechanism are moving slowly, others quickly. It is, however, easy to find out how much kinetic energy a body has if we know its weight and its velocity. Let there be a small ball hung from the point o, Fig. 26, by a silk thread, so that, when it vibrates, we can call it *a simple pendulum*. Now, you know that when it reaches the end of its swing at A it is, for a very short interval of time, motionless, and has no kinetic energy. It falls from A to B; and, as there is almost no friction, we may suppose that the potential energy which it loses in falling through the vertical height from A to B, is all stored up as kinetic energy when the ball reaches B. Now, suppose the body to have a certain velocity in feet per second when it reaches B. You know how to calculate* *the vertical height in feet through which a body must*

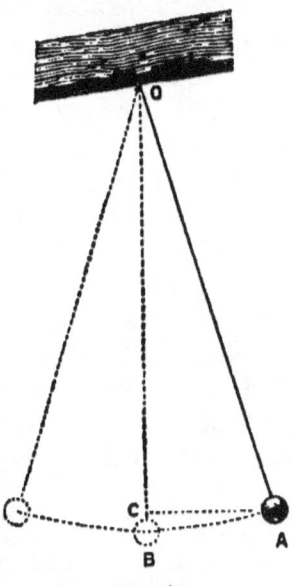

Fig. 26.

* *Laws of falling bodies.*—In the following rules v means the velo-

fall to acquire this velocity; it is the square of the velocity in feet ÷ 64·4. This is the vertical height from A to B. But the body has lost potential energy equal to its weight in pounds, multiplied by this height; and this is now stored up as kinetic energy. Hence, to find the kinetic energy of a moving body* *divide the weight of the body in pounds by* 64·4, *and multiply the quotient by the square of the velocity of the body in feet per second; the result will be the kinetic energy in foot-pounds.* In the case of the pendulum, this is the total energy of the bob. When the bob is at A all its energy is potential. When at B, all its energy is kinetic; and when it is anywhere between these positions, its total amount of energy is exactly the same as before, but part is potential and part is kinetic. During the swinging of a pendulum there is a constant change going on, potential energy changing into kinetic or kinetic into potential, and the sum of these two would always remain the same only that friction is constantly reducing this sum by converting part of it into energy of another order, namely, heat.

35. Test of the Law.—We now have a rule to find the energy stored up in a moving body, every part of which is moving with the same velocity. You can test this rule in the following way:—Get a

city of a body in feet per second, h is the height in feet from which the body has fallen, t is the time in seconds since the body began to fall; g is 32·2, and represents the effect of gravity in England.

$v^2 = 2gh$ { The square of the velocity of the body is found by multiplying the height by 64·4.

$h = \tfrac{1}{2}gt^2$ { Square the number of seconds during which the body has fallen, and multiply by 16·1; the product is the height fallen through in feet.

$v = gt$ { 32·2 times the number of seconds since the body began to fall, is the velocity of the body.

* The weight of a body in pounds, divided by 32·2, is the *mass* of the body; hence, the kinetic energy of a moving body is calculated by multiplying half its mass by the square of its velocity in feet per second. Kinetic energy = $\tfrac{1}{2}mv^2$.

pulley (Fig. 27) as light and frictionless as possible, because we must, at the beginning, neglect both the energy stored up in the pulley itself, and the loss by friction. Fasten the pulley at a considerable height above the floor. Let two equal weights, A and B, balance one another at the ends of a long, silk cord, passing over the pulley; and let there be a wooden scale, close alongside which A passes as it ascends and descends. Let us be able to fix to this scale, at any place, a plate which will suddenly stop A, and, above this, a ring which will just allow A to pass through. You will find such an arrangement as I speak of in almost every little collection of apparatus in the kingdom, and it is called an *Attwood's Machine.* Now, let A be as high as possible at the beginning; place on it a little weight, such as will be lifted off when A passes through a ring; and place a ring so that it will lift the little weight off A when A has fallen, say, 3 feet. You know that, so long as the little weight lies on A, the speed of A downwards and B upwards must become greater and greater. In fact, the potential energy lost by the little weight becomes converted into kinetic energy of the whole arrangement. Now, as soon as the little weight is stopped, A and B move with a steady motion; and if the table is placed by trial so that one second after A passes the ring it is suddenly stopped by the table, the distance between the ring and table shows the velocity which A, B, and the little weight had when the little weight was removed. In one experiment—A being 1 lb. and B the same, and the little weight 0·25 lb.—the velocity was measured after A had fallen 3 feet, and was found to be about 4·5 feet per

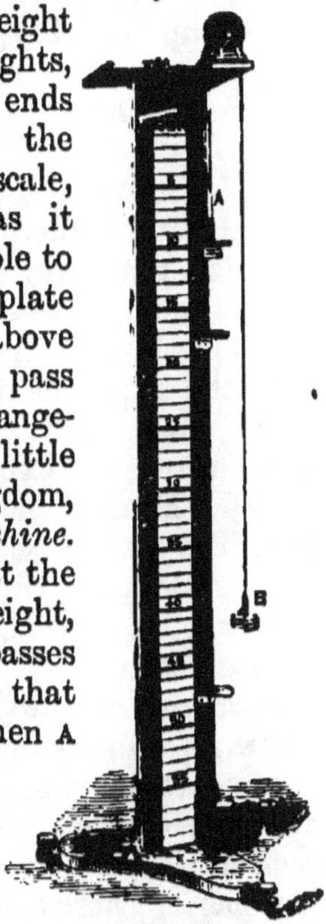

Fig. 27.

second. Now, the potential energy lost by the little weight was 3 × ·25, or ·75 foot-pound. The kinetic energy was stored up in 2·25 pounds, moving with the velocity of 4·5 feet per second; and, according to the above rule, its amount is

$$2\cdot 25 \div 64\cdot 4 \times 4\cdot 5 \times 4\cdot 5,$$

or ·71 foot-pound, or ·04 foot-pound too little. If we consider that there was some friction, that the pulley retained some kinetic energy, and that it was difficult to fix the table, so that exactly one second elapsed from A's passing through the ring until it was stopped, we see that the experiment is a fairly good illustration of the rule. You ought, with your own hands, to make a number of such experiments.

36. Energy in a Rotating Body.—Suppose now that the pulley is so massive that its kinetic energy is considerable, and may not be neglected, is there any way of finding from its speed how much energy it has stored up in it? We can easily calculate the energy in any little portion of a wheel if we know its velocity and mass, but those portions near the centre are moving more slowly than portions near the circumference, so that we have to calculate the energy in each little portion separately, and add all the results together. There is one thing which all portions of a wheel have in common—they all go round the centre the same number of times per minute. Suppose now that the number of revolutions of a wheel is doubled, the real velocity of every point in the wheel is doubled, whether that point be near the axis or not, so that the kinetic energy of the whole wheel is quadrupled; in fact, then, we find that the kinetic energy stored up in a wheel depends on the square of the number of revolutions which it makes per minute, so that, *the energy must be equal to a constant number multiplied by the square of the number of revolutions per minute.*

37. To find experimentally how much energy is possessed by a wheel when it is rotating, let the wheel be mounted on an axle supported on very frictionless bearings. If the centre of gravity of the wheel is not exactly in the axis, then it is better to place the wheel horizontally, as in Fig. 28. Now let

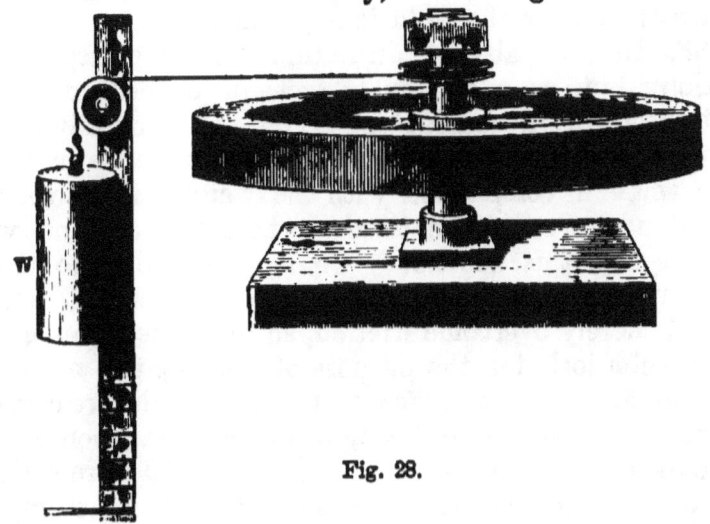

Fig. 28.

a cord wound round the axle be acted upon by a weight, w, which is only allowed to fall a certain distance. Suppose the weight to be 1,000 lbs., and that we only allow it to fall 8 feet from rest, so that when it has fallen this distance it no longer acts on the wheel, which will then rotate with a constant speed. Roughly speaking, the wheel possesses 1,000 × 8, or 8,000 foot-pounds of energy stored up in it. This is not quite true, because the weight itself possessed a certain amount of energy of motion which must be subtracted. Suppose that at the instant before being stopped the weight was moving with a velocity of 1·5 foot per second, then we must subtract

$\frac{1,000}{64\cdot 4} \times 1\cdot 5 \times 1\cdot 5$, or about 35 foot-pounds.

If there were no friction, and we find that a speed of 10 revolutions per minute has been given to the fly-wheel, we know that we have to find a constant number, **M**, which, when multiplied by the square of 10 or 100, will give 7,965 foot-pounds. Evidently **M** = 79·65, and hence, if ever we find this fly-wheel rotating, we know that it has stored up in it the amount of energy in foot-pounds 79·65 × square of number of revolutions per minute.

38. In the above calculation we have neglected friction; but, as a matter of fact, in experiments the **friction never is negligible.** You will generally find, however, that the weight which you use to cause motion is so small in comparison with the weight of the wheel that it adds very little to the whole friction, and we may neglect this addition. On a cord similar to that which you have already used, hang a small weight such as will merely overcome friction, so that when you give the wheel a jerk for the purpose of starting the motion, this weight will just suffice to prevent friction reducing the speed. Suppose this weight to be 5 lbs., then it is quite evident that 5 lbs. of the original 1,000 were really employed in overcoming friction and not in storage. Hence our calculation gives

995 × 8 − 35, or 7,925 foot-pounds as the total storage. This is at ten revolutions per minute. When it makes one revolution per minute the storage is 79·25 foot-pounds, and at any other speed we multiply 79·25 by the square of the number of revolutions per minute; 79·25 *is called the* **M** *of the wheel.*

39. It is obvious that you ought to be pretty quick in **counting the number of revolutions** of the wheel produced by the falling of the weight. Indeed, you ought to observe if possible the time taken in one revolution, using some special form of time-measurer, because the speed will now continually decrease on account of friction.

But there is another way in which it is easy to find the speed at the instant when the weight ceases to act. Find the total number of revolutions made by the wheel during the time that the weight is acting, and let some one observe this time in minutes. Then, as we know that the speed increases uniformly during this interval of time, the mean speed is just half the speed at the end of the interval; that is, *divide the number of revolutions by the number of minutes in which they were performed, and twice the quotient will give the number of revolutions per minute made by the wheel when the weight just ceases to act.* You can test your result by counting the number of revolutions from the time the weight is removed until the wheel is stopped by its own friction and dividing by the time which elapses; twice this quotient ought also to be the speed you want to know.

40. It is not necessary even to measure the friction directly, for we found that 7,965 foot-pounds were given out by the weight in falling; now *if we count the total number of revolutions made by the wheel from the time of starting until stopped by its own friction, and divide 7,965 by the total number, we shall find the loss of energy due to friction during one revolution,* since there is just as much energy wasted by friction in any one revolution as in any other. Ten times this must be the same amount of energy as 5 × 8, or 40 foot-pounds, for we measured the friction during 10 revolutions of the wheel as equivalent to 5 lbs. falling 8 feet. This, then, is the method you ought to employ.

41. You see that **M** is a number which ought to be known for every fly-wheel; it is just as important to know the **M** of a fly-wheel as to know the weight of an ordinary body. We have only to multiply the **M** by the square of the number of revolutions per minute, and we find at once the energy in foot-pounds stored up in the wheel. I have shown you how to find the **M** of a fly-wheel by experiment; I will now give you an idea of its value in different cases. I cannot prove to you this first rule, or the other rules given in the foot-note. Imagine a **grindstone** whose diameter is 4·5 feet, whose breadth is 1·4 foot, the weight of its material per cubic foot being 132 lbs.; then we can calculate its **M** by first finding

$$132 \times 1\cdot 4 \times 4\cdot 5 \times 4\cdot 5 \times 4\cdot 5 \times 4\cdot 5,$$

and dividing this answer by 29,900. For any rotating object of cylindrical shape, the shape of a grindstone, this rule will always find **M**. *Multiply the weight of the material per cubic foot by the breadth or width; multiply this by the fourth power of the diameter, and divide by the constant number* 29,900. Whether the material is wood or stone or metal, this will give **M**, and this multiplied by the square of the number of revolutions per minute will give the energy in foot-pounds stored up in the rotating body. For the above grindstone, on calculating out, you will find the **M** to be 2·53. So that when it makes 1 revolution per minute, there is stored up in it 2·53 foot-pounds of energy; when it makes 2 revolutions per minute, there is stored up in it 2·53 × 4, or 10·12 foot-pounds

							Foot-pounds.
At	3 revolutions,	2·53 ×		9,	or		22·68
,,	20	,,	2·53 ×	400,	,,		1,012
,,	50	,,	2·53 ×	2,500,	,,		6,325
,,	100	,,	2·53 ×	10,000,	,,		25,300

42. The energy stored up in a rotating body is equal to $\frac{1}{2} I a^2$, where I is **moment of inertia** about the axis; that is, the sum of all such terms as mass of a little portion multiplied by the square of its distance from the axis. a is angular velocity (see GLOSSARY) in radians. Hence, as $a = \frac{2n\pi}{60}$, if n is number of revolutions per minute, and π is 3·1416, the energy is $I \frac{n^2 \pi^2}{1800}$, so that our **M** is $\frac{I \cdot \pi^2}{1800}$. For the following bodies I give in Table I. the values of I and **M**. w is weight in lbs. per cubic foot, and dimensions are in feet.

43. If we fix a small weight of 20 lbs. on a wheel, at 12 feet from the axis, this adds to the **M** of the wheel the amount

$$20 \times 12 \times 12 \div 5,873, \text{ or } 0·49;$$

or the weight multiplied by the square of its distance from the axis, divided by 5,873.

If we add a very *thin* rim to a wheel, the addition to **M** is found by multiplying the weight of the rim by the square of its average radius, and dividing by 5,873; or,

TABLE I.

	Nature of Rotating Body.	I	M
	Sphere of diameter d, rotating about diameter as axis	$wd^5 \times \cdot 001626$	$wd^5 \div 112{,}166$
	Spherical shell, whose outside diameter is d, and inside is d_1, rotating about diameter as axis	$w(d^5-d_1^5) \times \cdot 001626$	$w(d^5-d_1^5) \div 112{,}166$
	Cylinder, diameter d, length l, rotating about its axis.	$wld^4 \times \cdot 00305$	$wld^4 \div 59{,}814$
	Hollow cylinder, outside diameter d, inside diameter d_1, length l .	$wl(d^4-d_1^4) \times \cdot 00305$	$wl(d^4-d_1^4) \div 59{,}814$
	Thin rim, mean radius r of weight w . . .	$wr^2 \div 32 \cdot 2$	$wr^2 \div 5{,}873$
	Thin rod, of length l, rotating about axis through its middle point, at right angles to its length. Weight of rod w	$wl^2 \cdot 00258$	$wl^2 \div 70{,}474$
	Thin rectangular plate, rotating about axis through its centre parallel to side b, the side d being at right angles to axis. Weight of plate w	$wl^2 \cdot 00258$	$wl^2 \div 70{,}474$

multiplying the weight of the rim by the square of its average diameter, and dividing by 23,492.

It will be found that if a fly-wheel has light arms and a heavy rim, as we often see on such wheels, *a fairly good approximation to its* **M** *is found by multiplying the weight of the rim by the square of the mean diameter of the rim, and dividing by* 23,000.

Example.—The rim of a fly-wheel weighs 15 tons; its mean diameter is 20 feet. Calculate approximately what energy is stored up in it when it makes 60 revolutions per minute. Here you will find the **M** of the fly-wheel to be about 584, and hence the stored energy is $584 \times 60 \times 60$, or 2,102,400 foot-pounds.

44. Steadiness of Machines.—A fly-wheel is put upon a riveting or shearing machine, or other machine, because the supply of energy to the machine is not given regularly, or else because the demand for energy from the machine is irregular. The fly-wheel enables the machine to maintain a more constant speed. In calculating the proper size of a fly-wheel for any machine we must know two things. First, what is the greatest alteration of speed allowable in the case; and secondly, the greatest fluctuation of the demand and supply of energy. Thus, suppose we wish never to have the speed of the fly-wheel more than fifty-one nor less than forty-nine revolutions per minute, and that during some interval of time the fly-wheel has to give out 20,000 foot-pounds more than it receives during that time; then, although the fly-wheel will afterwards have this deficiency made up to it by some steady supply, it is obvious that its speed must diminish. We wish its speed to diminish only from fifty-one revolutions to forty-nine revolutions per minute in this interval of time. Now, when the fly-wheel runs at fifty-one revolutions, it has stored up an amount of energy equal to its $\mathbf{M} \times 51 \times 51$; and when it runs at forty-nine revolutions, its store is $\mathbf{M} \times 49 \times 49$, and the difference between these two ought to be 20,000. Hence, subtracting 49×49, or 2 401, from 51×51, or

2,601, we get 200; and dividing 20,000 by 200, we find 100 as the required value for **M**. *Subtract, then, the square of the least speed from the square of the greatest, and divide the greatest excess of demand or supply by this remainder; the quotient is the* **M** *of the fly-wheel.* Having found **M**, the question is, how can you tell from it the size and weight of the wheel? Find the **M** of any wheel of the same shape and material as that which you want to use. It is obvious that *the diameters of the wheels are as the fifth roots of their* **M**'s.* We want a wheel whose **M** is 100. Suppose I find a wheel of the shape I wish to use whose outer diameter is 8 feet, and I calculate its **M**, and find it to be 11; then

The fifth root of 11 : fifth root of 100 :: 8 : answer.

Log. 11 = 1·0413927; divided by 5 it is 0·2082785, which is the logarithm of 1·615.

Log. 100 = 2·0; divided by 5 it is 0·4, which is the logarithm of 2·512. Hence

1·615 : 2·512 :: 8 : answer.

This is an easy exercise in simple proportion. I find my answer to be 12·44 feet, or 12 feet 5¼ inches, the diameter of the required fly-wheel, which is to be similar in form to the smaller specimen used by me for calculation.

* If we have any two similar wheels, or other rotating bodies of the same material; if we consider any similar small portions of them; it is evident that their weights are proportional to their cubic contents, or to the cubes of any similar linear measurements. Hence, if one is, say, twice the diameter of the other, as every dimension of the one is twice that of the other, the weight of one must be 2 × 2 × 2, or eight times that of the other. Now, the **M** of any rotating body depends, not merely on the weight of each portion of the body, but on the square of its distance from the axis, so that the **M** of one must be 8 × 2 × 2, or thirty-two times the **M** of the other. Similarly, if the linear dimensions were as 3 to 1, the values of **M** would be as 243 to 1 for a pair of similar wheels.

Example: We want a wheel which will have a store of 1,000 foot-pounds when rotating at twenty revolutions per minute, and it is to be of the same shape as that of an already existing wheel, which is four feet in diameter, and which contains a store of 1,350 foot-pounds when running at thirty revolutions. Evidently the **M** of this second wheel is 1,350 ÷ 900, or 1·5, and the **M** of the first wheel is to be 2·5. Using logarithms, we find that the fifth root of 1·5 is to the fifth root of 2·5 as 4 feet is to 4·4 feet, the answer.

E

45. The total kinetic energy stored up in any machine is found by calculating the energy in every wheel and in every moving part, and adding all together. But suppose that in the machine there is some shaft of more importance than any other, it is usual to give the speed of this shaft only, because if its speed be doubled, the speed of every other is doubled. Thus, in a steam-engine we state the number of revolutions per minute of the crank shaft, and this tells us the speed of every part of the engine. Let, then, the number of revolutions of some such principal axle of a machine be found. If this number of revolutions is doubled, the kinetic energy stored up in the machine is quadrupled; and, in fact, *the kinetic energy stored up is equal to a certain number which can be found for the machine, and which we shall call its* **M**, *multiplied by the square of the number of revolutions of this particular axle per minute.* The **M** of *any* machine may be experimentally determined in exactly the same way as we have shown above.

If we know the **M** of any machine, then the **M** of any other machine made to the same drawings, and of the same materials, but with all its dimensions twice as great, is thirty-two times as great, because the **M**'s of the two machines are proportional to the fifth powers of their corresponding dimensions.

46. Facts Useful to Know.—In a condensing steam-engine, when the steam is cut off at from one-third to one-eighth of the stroke, there is a certain portion of the stroke during which ·16 to ·19 of the total work done during the stroke is given out by the steam to the engine, in excess of that given out by the engine itself as useful work. In a non-condensing engine, steam cut off at from one-half to one-fifth of the stroke, the excess work is ·16 to ·23 of the total work of one stroke. These facts will enable you to calculate the proper size of fly-wheel for a given steam-engine when you know the work done in one stroke, and also the greatest and least speeds allowable. For a punching, shearing, or riveting machine no figures of this kind are available. Observations are much needed. In the case of a pump with a fly-wheel it is easy to calculate the excess work done during any period. But in many kinds of pump great variations of speed are generally allowed.

CHAPTER VI.
EXTENSION AND COMPRESSION.

47. How a pull is exerted.—How is it that a cord transmits force from my hand to an object when I pull the object by means of a string? If you study this matter you will see that every particle of the string coheres to the next, and although the refusal of one particle to come away from its neighbour might easily be overcome, there are so many of them to be separated at any particular section of the string that it requires a considerable pull to perform this operation. When a string is pulled it really lengthens a little, and it lengthens more the more force is applied, although it may not break. A string is not so easy to experiment with as a wire of metal, because we find that it differs more in its quality at different sections, and it is affected by dampness and many other circumstances. No doubt it is also difficult to obtain a metal wire which shall just be as willing to break at one place as another, that is, which shall be exactly of the same material everywhere; but metal wire is certainly more to be depended upon than string.

48. Strain.—Take, then, a steel wire, A B (Fig. 29), fastened near the ceiling at A, between two pieces of wood, screwed together firmly so that there may be no tendency for the wire to break just at the fastening. Similarly fasten at B a scale-pan arrangement, and, first, place just so much weight in the pan as keeps the wire

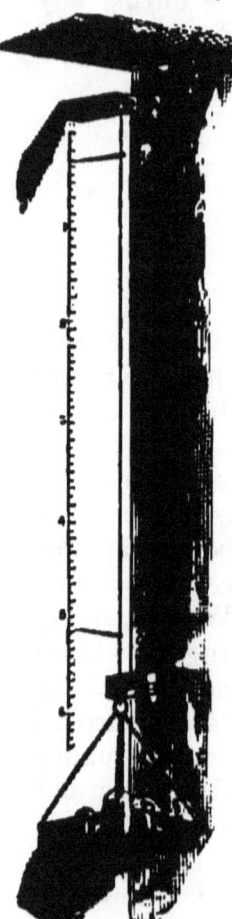

Fig. 20.

taut. Let there be two light little pointers stuck or tied on at *a* and *b*, and let there be a vertical scale on the wall. Now read off the distance between *a* and *b* on the scale and note the weight. Add more weight, and again read the distance, and continue doing this until the wire breaks. You will prove by means of squared paper that—

The amount of the extension of a wire is proportional to the weight which produces the extension.

When we speak of the *strain* in the wire, and want to use the term *strain* in an exact sense, we mean *the fraction of itself by which a b lengthens*. Thus, suppose that *a b* was 50 feet, and that it lengthens 1 foot, we say that the strain is $\frac{1}{50}$, or ·02, or 2 per cent. I need hardly tell you how important it is to learn the exact meaning of a word like this; it will give clearness to your ideas.

49. Stress.—If you take another wire of the same material, but of twice the sectional area of this one, you will find that it needs twice as much load to produce the same strain. The reason of this is that you have at any section twice as many particles of steel resisting the pull. The pull produced by the load acts at every cross section in the same way, no matter how long the wire may be; but if the wire is thicker at one place than another, then at such a cross section the pull is distributed over a greater number of pairs of particles. We see, then, that if a wire or rod is transmitting a pull, it is well not to consider the total load, but rather the load per square inch of section. *The load per square inch is called the stress.*

This is the exact meaning which we give to the word *stress*. Much of the difficulty you may have met with in your reading is due to the fact that you have not made a proper distinction between the meanings of these two words. *Stress* is the load per square inch which produces a fractional alteration of the length of a wire or rod, and this fractional alteration is called the *strain*. Suppose your load to be 6 lbs., and your wire circular in section, with a diameter of 0·05 inch. Then the area of the section is 0·025 × 0·025 × 3·1416, or ·00196 square inch. The stress is 6 ÷ ·00196, or 3,061 lbs. per square inch. You will find that this thin wire gets the same strain with a total load of 6 lbs. as a rod one square inch in section would get with a load of 3,061 lbs. If ever you get a problem to work out, relating to the lengthening of a wire or rod produced by a load, you must consider, not the total lengthening of the wire or rod, but its fractional amount of lengthening, and call this the *strain;* also consider, not the total load, but the load per square inch of section, and call this the *stress*, and you will find that for some kinds of wrought iron—

The stress = the strain × 29,000,000.

Example.—How much extension is produced in a wrought-iron tie-rod 80 feet long, whose cross section is 3 square inches, by a pull of 9,000 lbs.? Here the stress is 3,000 lbs. per square inch, and 3,000 is equal to 29,000,000 times the strain, or 3,000 ÷ 29,000,000, or ·0001034 is the strain. The extension is the fraction ·0001034 of the total length, and 80 feet × ·0001034, or ·00828 foot is the answer—nearly one one-hundredth of a foot, or, more nearly, the tenth of an inch.

50. It is somewhat more difficult to experiment on the shortening of a strut or column when it transmits a push, because you cannot use very long struts. A strut, as you know, tends to bend if it is very long; and when it breaks, unless great care is taken to keep it straight, it breaks more easily the longer it is. The bending action causes the load to act more on one part of the cross section than another, and the stress—or the pushing force—per square inch is greater at one part of the section than at another. If you experiment, therefore, you must take care to use struts which are in no danger of bending. In Chap. X. I shall consider the bending of beams, after which you will better understand the present difficulty. It is sufficient for you at present to know that, **whereas the pull in a tie-bar tends to make it straighter if possible, the push in a strut tends to make it bend.** Hence, *in an iron railway-bridge or roof you will see that the tie-bars are thin solid rods usually, and they might be chains or ropes if these were cheap enough; but the struts must not merely have a proper area of cross section, this cross section must also be wide in every direction.* Thus, instead of a solid cast-iron column you always see a hollow one, unless the column is very short. Also, a thin plate of iron suffices for the lower boom or flange of a railway-girder (because it resists a pull), whereas the top boom is a hollow tube, or is U-, or ∩-, or ⌐⌐-shaped, because it must resist a push. Long struts, therefore, must be considered in Chap. XII., after we have investigated the bending of beams.

51. A Short Strut will be found to obey exactly the same laws as a tie bar. The load per square inch is called the *stress*. The *shortening is a fraction of the whole length of the strut, and this fraction is called the strain.* You will find from your experiments that the strain is proportional to the stress. Thus for wrought iron struts or columns

The stress = the strain × 29,000,000.

The multiplying number is found to be the same for the same material, whether it resists a push or a pull. This number is called "Young's Modulus of Elasticity;" it has been measured for various materials, and is given in Table III. In using it you must remember that the stress is in pounds per square inch.

LIMITS OF ELASTICITY.

Exercise 1.—By how much would a round bar of steel, 120 feet long, whose diameter is 2 inches, lengthen with a pull of 30 tons ? Answer : 0·0855 foot.

Exercise 2.—By how much would a column of oak, 7 feet long and 4 inches square, be compressed in supporting a weight of 2 tons ? Answer : 0·0013 foot.

52. I have said that if you use squared paper after making your experiments, you will find that the strain is proportional to the stress, and the lengthening of a tie bar is proportional to the total pulling force. But you will find that this law is not true when the loads become too great. If your loads are less than a quarter of the breaking load, you will find on removing them that the wire on which you are experimenting goes back to its original length.* But if your loads much exceed this amount, it will be found that the wire has taken a **permanent set**; that is, if you remove the load the wire will not go back to its original length. It remains permanently longer than it originally was, and we say that we have exceeded the *limits of elasticity*. The load which produces this permanent set is said to be the measure of the **elastic strength** of the wire, for although it does not break the wire it alters it permanently. Now, it is only for loads less than this that the law " strain is proportional to stress," is true. Your squared paper for experiments on a steel wire would give a straight line becoming a curve, like Fig. 30.

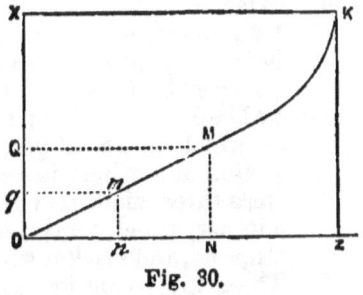

Fig. 30.

---

* It may not go back at once to its old length, but in a few minutes it will be found exactly where it was before you loaded it. Similarly, when the load is put on, there is first a sudden lengthening and after this there is a slight extension going on so long as the load remains, but it practically comes to an end in a few minutes. This after-action is so slight that I have not till now spoken about it, although we have reason to believe that its investigation would be of great importance.

When you plot your results, making the distance $m\,n$ represent the extension of the wire for a load represented by the distance $m\,q$, to any scale you please, you will find that the line passing through your points is straight only from O to M say, and then it curves upwards. The distance, Q M, represents the load which produces permanent set. For greater loads than this, the extension is more than proportional to the load, and increases more rapidly until we get at K a very rapid extension indeed, for the wire broke with the load, K X, and just before it broke its extension was K Z.*

53. **The nature of the strain in a wire** which is being extended, or in a column which is being compressed, cannot be said to be simple. If all lines in one direction, and in one direction only, became shorter or longer the strain would be called *simple*, but it needs rather a complicated system of external pressure to produce this effect. No matter how a body is strained, if we consider a small portion of it we shall find that any strain simply consists of extensions and compressions in different directions. In fact, imagine a very small spherical portion of the body before it is strained; the effect of strain is to convert the little sphere into a figure called an ellipsoid, that is, a figure every section of which is an ellipse, or a circle; remember that every section of a sphere is a circle. It may be proved that there were three diameters of the sphere at right angles to one another, which remain at right angles to one another in the ellipsoid, and are known as *the principal axes* of the ellipsoid. These directions are now called the *principal axes* of the strain existing at that part of the strained body. Along one of these directions the contraction (or extension) is less, and in another greater, than in any other direction whatever.

54. **Example.**—Thus if M′ N′ (Fig. 31) is part of a long wire subjected to a pull, the portion of matter which was enclosed in the very small imaginary spherical surface, A B C D,

---

* Instruments have been designed which register on a sheet of paper (as the pencil of a steam-engine indicator does) the load pulling a rod, and the extension which it produces. A little brass cylinder covered with paper is touched by a pencil on the end of the rod. The amount of rotation of the barrel is regulated so that it is proportional to the load. By this means, curves, like that of Fig. 30, may rapidly be drawn as the load on the rod is gradually made to increase till the rod breaks. (See Art. 125.)

before the pull was applied is now enclosed in the ellipsoidal spherical surface, A′ C′ B′ D′. The sphere has become an ellipsoid of revolution; A B becomes A′ B′, C D becomes C′ D′. The strain in the direction A B is $\frac{A'B' - AB}{AB}$ and this is equal to the pull in the wire per square inch divided by Young's Modulus of Elasticity, E. As, however, it is often more convenient to use a multiplier than a divisor we are in the habit of using the reciprocal of E, and denoting it by the letter a. Thus, if the pull per square inch is one pound it produces a strain of the amount, a, in the direction A B; the lateral contraction of the material is $\frac{CD - C'D'}{CD}$ and in this case is usually denoted by the letter b.

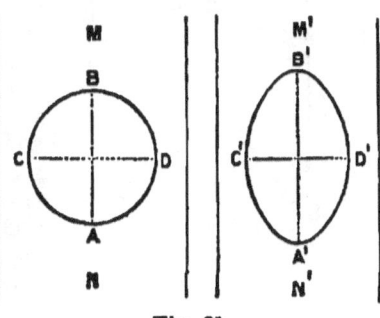

Fig. 31.

55. The diminution in bulk of a substance when it is subjected to pressure uniform all round, as, for instance, when it is surrounded by water in a hydraulic press, or sunk in the sea, has been experimented upon. The lessening in the bulk per cubic inch is called the cubical strain of the substance. The pressure in pounds per square inch all over its surface represents the stress, and it is found that the strain is proportional to the stress. In fact, in any substance the stress is equal to the strain multiplied by a certain number, for which the letter K is usually employed, called the *Modulus of Elasticity of bulk*.

TABLE II.

Substance.	Modulus of Elasticity of Bulk in pounds per square inch.
Ether	120 thousand.
Cold Water . . .	300 ,,
Water at 130° Fahr. .	330 ,,
Mercury	5 millions.
Flint glass . . .	6 ,,
Cast iron	14 ,,
Wrought iron . .	20 ,,
Copper	24 ,,
Steel	30 ,,

Imagine a cube one inch in each edge (Fig. 32), subjected

to a uniform compressive force of 1 lb. per square inch on the opposite faces, A D E F and B C L G. Evidently the edges A B, C D, L E, and G F, become $1 - a$ inch in length, a being the reciprocal of Young's Modulus used above. Also the edges A D, B C, G L, and F E, get the length $1 + b$ inch. If now we give to the faces, A B C D, and E F G L, of this cube, compressive forces 1 lb. per square inch, it is the edges A F, &c., which shorten, and the edges A B, &c., which lengthen. Again, give the compressive forces to the third pair of opposite faces, A B G F and C D E L, and we have the edges A D, &c., shortening and B G, &c., lengthening. If, now, all three sets of compressive forces act at the same time, that is, the cube gets on every face a pressure of 1 lb. per square inch, as the compressions and extensions are exceedingly small, each edge shortens by the amount a and lengthens by the amount $2 b$. Hence the edge which used to be 1 inch is now $1 - a + 2b$ inch. The cubic contents used to be 1 cubic inch, it is now $1 - 3 (a - 2 b)$ with great exactitude. Hence $3 (a - 2 b)$ is the amount of cubical strain produced by 1 lb. per square inch. That is, the *Modulus of Elasticity of bulk*,

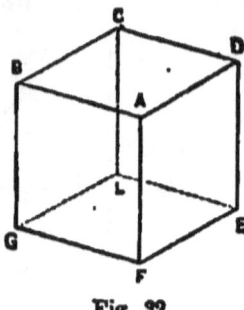

Fig. 32.

$$K = \frac{1}{3(a - 2b)}$$

and if we know a and b it may be calculated.

56. It is found then that when a rod is pulled, **not only does it get longer, but its diameter gets less.** When, for example, a rod of glass is pulled so that its length increases by the one-thousandth of itself; it is found that its diameter gets less by the one three-thousandth of itself.

57. Strength.—Table III., p. 68, shows among other things the pulling (or tensile) and pushing (or compressive) stress which a material will bear before breaking. Probably if these stresses were allowed to act on the material for some time it would break even if they were not added to. They are obtained from experiments in which the load was increased pretty quickly, and yet quietly, that is, without any jerking or sudden action. The numbers in the table are taken from many sources, and must in

STRENGTH OF PIPES AND BOILERS.

general only be regarded as giving rough average values. The strengths of the metals I have taken from Mr. Unwin's book on machine design. *The stress per square inch which will produce a permanent set in the material is sometimes called the elastic strength. The working stress is usually a fraction of this;* it is the stress which experience tells us to calculate upon for loads acting for a long time on materials, and which we shall be sure are perfectly safe in the case of such materials as are supplied from foundries and forges.

Exercise 1.—How great a pull will a round rod of brass stand before it breaks, if its diameter is 0·3 inch? What pull would produce in it a permanent set, and what is the safe working pull? Answers: 1,237, 484, and 254 lbs.

Exercise 2.—A short hollow cylindric column of cast iron is 8 inches in outer diameter, 5 inches inner diameter. What is the safe load and what load will produce permanent set? Answer: the area of cross section is 4 × 4 × 3·1416 minus 2·5 × 2·5 × 3·1416, or 30·63 square inches; 30·63 × 21,000 is 643,230 lbs., or 287 tons; 30·63 × 10,400 is 318,522 lbs., or 142 tons.

58. Pipes and Boilers.—We may consider that a pipe or other hollow cylinder, when it tends to burst with internal pressure, has twice as much tendency to burst laterally as to burst longitudinally. If, however, the cylinder is short, the ends may modify this effect, strengthening the cylinder laterally without altering the endlong strength, but it is usual to have such cylinders long in proportion to their diameter, hence it is their lateral strength which has to be considered.

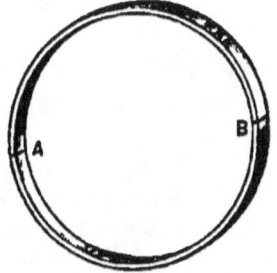

Fig. 33.

Imagine a hoop of breadth one inch, the pressure per square inch inside multiplied by the diameter in inches is the total force which tends to make this hoop break at A and B (Fig. 33), or at the ends of any other diameter. The tendency to burst is resisted by the tension at A and B, so that the area in square inches at A, together with that at B, is like the area of cross section of a tie-rod subjected to a total pull of the above amount. Hence we have the rule, **the greatest safe pressure per square inch inside a boiler or pipe, multiplied by the diameter (in inches), is equal to twice the thickness of the metal multiplied by the safe working tensile stress of the material per square inch.** It is in this way that we calculate the strength of a boiler or large water-pipe. When the boiler has riveted joints we must, of course, regard the material as weaker than if it could resist tensile stress everywhere like a continuous boiler plate. In cast iron pipes and in steam-engine cylinders it has to be remembered that the difficulty in getting castings which are of the same thickness everywhere, and the allowance that must be made for tendency to cross-breaking when the pipes are handled, as well as the great allowance that must be made in steam-engine cylinders for stiffness, the difficulty of casting, and boring out, cause such calculations as the above to be somewhat useless. Thus it will usually be found that, whereas a large cast iron water-pipe is not much thicker than the above calculation would lead us to expect, yet a thin cast iron pipe is often of more than twice such a thickness.

59. It is easy to prove the truth of the statement made at the beginning of the last paragraph—a boiler has twice as much tendency to burst laterally as longitudinally.

When a boiler bursts endwise, the area of section at which fracture occurs is the circumference multiplied by the thickness, or $3·1416\, d\, t$ if d is the diameter and t the thickness. But the total endlong pressure is the pressure on the sectional area of the boiler at the place, or $·7854\, d^2 p$, if p is the fluid pressure in pounds per square inch, hence $·7854\, d^2 p \div 3·1416\, d\, t$, or $d\, p \div 4\, t$ is the stress on the material.

Chap. VI.] STRESS ON A SPHERICAL BOILER. 61

In the lateral bursting tendency considered above, the total force acting at A and B (Fig. 33) is dp pounds, and the area of the metal at A and B being $2t$ square inches, the stress on the material is $dp \div 2t$, or twice as much as in the other case.

In this investigation I have considered that the stress at A and B is one of mere tension, and this is the case when the metal is thin in comparison with its diameter. In a thick pipe or in a gun it is found that, although the average stress may be arrived at in the above way, some portions of the metal are more severely strained. It is not my purpose to consider such cases in this book.

In a spherical boiler the tensile stress anywhere is evidently $dp \div 4t$, being the same as the endlong stress in a cylindrical boiler of the same diameter and thickness.

Some students may have difficulty in understanding how it is that the **pressure tending to burst a spherical boiler** at A B is found by multiplying the cross-sectional area at A B by the pressure of the fluid in pounds per square inch. The pressure is really exerted on every portion of the surface A C B (Fig. 33A), and it is everywhere at right angles to the surface, but if the resultant force is calculated it will be found to be what has been stated. This will become evident on considering that if one half of such a boiler is closed by a flat plate at A B (Fig. 33B), as the fluid pressure does not cause motion of any kind, its resultant action on any portion of the surface must be equal and opposite to its resultant action on all the rest of the surface, therefore the resultant pressure on A C B is equal to the pressure on A B. It is for this reason that we always calculate the pressure on a pump plunger as being the sectional area in inches of the plunger multiplied by the pressure per square inch of the fluid, taking no account of the fact that the end of the plunger may either be rounded or flat.

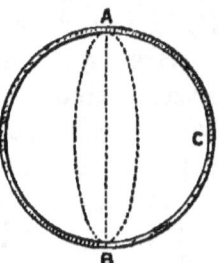

Fig 33A.

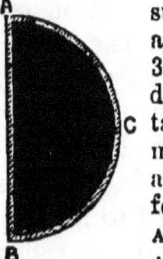

Fig. 33B.

CHAPTER VII.
PECULIAR BEHAVIOUR OF MATERIALS.

60. In this chapter I wish to draw your attention to a subject in which **the workman** is more likely to obtain valuable information than any other person. I have told you that when a load continues to pull a wire the wire continues to lengthen, although for small loads the extension is practically ended after a few minutes. A load which is so great that it strains the wire permanently will very often be quite unable to break the wire, however long it is applied, but it is never thought advisable to allow such a load to act for a long time. It is found that after getting such a permanent set a wire is more elastic, that is, its **elastic strength** is greater than it was previously. A man who puts up bells in a house "kills" his copper wire, that is, gives it a permanent set, as he finds that, after this operation, it **obeys** better the laws of elasticity referred to above. Similarly a telegraph-line man kills his iron wire before fixing it to the telegraph posts. It is probable that the effect of this "killing" is like the straining of a piece of **riveted work** beyond its limits of elasticity, which makes all the rivets fit better into their beds. It is, however, very curious to see how much set can be given to some materials. For instance, a thin brass wire gently pulled may be twisted to an enormous extent, and still retain elastic properties; indeed, its elastic strength may be higher than it was in the beginning. Again, when a wire by being drawn through a die is reduced to a smaller size, there is a complete alteration in the arrangement of its particles, and yet we know that the drawn wire has usually greater strength than it had originally, that is, it will bear a greater load per square inch of its section; even hardened steel wire can be drawn in this way. In the same way, a block of copper, by a series of beatings and temperings, may be shaped like a pot or boiler, and a coin take the impression of a die, without losing their strength. In fact, metals seem to be able to flow if sufficient stress is applied to them, and at the end of the operation they are as strong as ever; indeed, they are very often stronger than before. When they are a little harder than they were, this quality, if not wanted, can be removed by heating and slow cooling, a process which goes by the name of "annealing."

61. Another noticeable fact. It is found that a watch goes faster and faster for some time after it is made, but at the end of some months the balance spring settles down into a state which does not much change afterwards. In this state then its elasticity is greater than it was in the beginning. The springs of chronometers are, however, often laid aside as useless after a few years' service, their elastic condition having altered so much since the beginning that they have to be replaced. It has been found that when a long wire is kept slightly twisting and untwisting except on Sundays, there is a gradual softening or an increase of internal friction going on all the week, which greatly disappears during the Sunday rest. This and other facts concerning the behaviour of materials which have been overstrained are vaguely comprehended under the expression, "fatigue of materials."*

62. When a load is suddenly applied to stretch a wire, it produces greater effects than when slowly and quietly applied. We know the reason of this. A weight which slowly applied would produce an extension of one inch, would, when suddenly applied, produce an extension of two inches. The wire now shortens to its original length; then extends two inches and continues to get shorter and longer alternately. As

* Consult Sir Wm. Thomson's article on *Elasticity*, in the "Encyclopædia Britannica."

there is friction of some kind among the particles of the wire, and there is also external friction, the lengthenings and shortenings gradually lessen till, in a short time, the wire settles down into the same state as it would have been in if the load had been slowly applied. Now, if we suppose this wire, when stretched two inches, to be strained just beyond its elastic strength, it is evident that the suddenly applied load does harm, whereas, the same load slowly applied would do no harm. The harm is greater if the weight, besides being applied suddenly, is moving before it begins to act on the wire. Take the case of a stone which is being removed by means of a crane. If the stone, happening to fall a little, be brought up by the chain, the increase in the stress on the chain is simply proportional to the height from which the stone has fallen, and is greater the less the chain is extended (see Chap. XIX.). When a wire is lengthened ·1 foot by a weight of 1,000 lbs., which has been increased gradually, we know that the pull on the wire began with 0, and, as the wire gradually extended, the pull became greater, till it is now 1,000 lbs. The average pull was 500 lbs. and 500 × ·1, or 50 foot-pounds is the total energy stored up in the wire in the shape of a strain. If we wish to give more energy to the wire, we must strain it more; and this is just what we do when we let the weight fall suddenly.

63. **The energy stored up in any strained body** may be calculated if we know the stress and the strain. The main-spring of a watch contains a store of energy which is gradually given out by the spring in returning to an unstrained condition. Each strained portion of the spring contains a portion of the store, and if at any place in the body there is too great a store, the body will break there. Let us consider why a chisel cuts into an iron plate. When I strike the head of a chisel with a hammer I give to the chisel in a very short period of time a certain amount of energy. This energy is transmitted very quickly to the plate through the edge of the chisel. The shorter and more rigid the chisel, the more quickly is the energy sent through the cutting edge into a portion of the

plate. If it is not conveyed away rapidly from the edge, the amount contained in a small portion of material just under the edge is very great, and the material is fractured there. As the energy of strain is proportional to the product of stress and strain or to strain squared, the **possibility of fracture** for a material is represented by the **square root of the strain energy** it contains per cubic inch. If a material is brittle there is a sort of instability which causes fracture at one place to extend to all neighbouring places. And hence, if we deliver with great rapidity to a small portion of such a material a moderate supply of energy, it is sufficient to produce a large fracture. As our material becomes less and less brittle, we must have, over a larger and larger part of the volume in which we want fracture to occur, a sufficient supply of strain energy delivered. Hence, in cutting wood we use a wooden mallet and a more or less lengthened wooden-headed chisel. The mallet and chisel act as a reservoir for the energy of the blow which is delivered to the wood from the edge of the chisel with comparative slowness and just in sufficient quantity to cause rupture in front of the edge. If the wood without gaining in strength became **more rigid** so as to be able to **carry off more rapidly** the energy given to it by the chisel's edge, it would be necessary to make the supply more rapid by using a more rigid chisel and mallet, and as we do this we must take care that the chisel itself near the edge is strong enough to resist fracture (see Chap. XIX.).

64. These are facts which we can understand; the following, however, are not so self-evident. A piston rod is subjected to **tensile and compressive stresses, often repeated.** It is found that its breaking strength is not 45,000 lbs. per square inch, which, let us say, it would be for a steady pull or push, but 15,000 lbs. per square inch. If, instead of such an action, we have a tensile stress which varies frequently, although not suddenly, from 30,000 lbs. per square inch to zero, the rod will break after a time. In the same way, steel which will bear a steady stress of 84,600 lbs. per square inch, will only bear 46,500 lbs. per square inch if the stress varies between this and zero, but is always of the same kind; whereas, it will only bear 25,000 lbs. per square inch if the stress is sometimes a pull of

F

this amount, and is sometimes a push of the same amount.

65. We are also quite ignorant of the reason **why steel hardens when suddenly cooled**, and why this hardness is different according to the temperature from which this cooling starts. In every workshop the common method adopted for **tempering a fitter's chisel** is as follows :—Heat the chisel to a dull red colour, put the edge in water to a distance of say half an inch, quickly rub with pumice or a file, watch the edge till, as it heats by conduction from the thicker portion you know that a certain temperature has been reached by seeing a certain colour (lightish yellow for a chisel) of oxide of iron making its appearance. When this colour appears plunge the whole chisel into water. The steel is first made extremely hard at its edge, and is then brought back to the required degree of hardness by re-heating up to a certain temperature and then suddenly cooling. This simple process is in common use. In tempering other objects sometimes much greater care must be taken, since it is often necessary that every portion of the object shall be of the same hardness, and in such cases the whole may be cooled at first and then reheated in a bath of oil, mercury, or other melted metal whose temperature is definitely known. The effect is of the same kind, however, whether the process is the rough one which I have described or a more careful one. It is usual to explain it by saying that in sudden cooling the particles of steel have not had time to get into their natural positions when cold, and that they jam each other somehow, getting into positions of instability ; but if it be remembered that we often find steel when hard to be stronger than when it was soft, you will see that there is a great deal wanting in this explanation of what occurs. As regards the influence of impurities, of gases from the atmosphere which are suddenly imprisoned among the particles of steel, very little is yet known. Again, **cast iron** is stronger

if compressed in the melted condition until it solidifies, and we explain this vaguely by saying that the pressure closes up little cavities. **Metal wires are** strengthened in being drawn smaller through dies, but they lose this increase of strength, and gain in toughness, when afterwards heated and cooled slowly.

66. I need not give you any more items of a long catalogue of curious properties of materials which we do not yet understand. **Workmen know of** and depend upon many of these actions, but nobody seems to have any clear idea as to how they take place. It is not merely that workmen temper steel and find that curious changes occur in the properties of their steel when it is altered a little in its chemical state; the philosopher and the workman are equally aware of these facts, and equally ignorant of their real nature; but some workmen who deal with little mechanical contrivances make use in their trades of certain properties of brass and iron and steel which the philosopher is quite ignorant of, and it is possible that an observing workman who knows a little of chemistry and physics may discover the key to all the mass of hitherto unexplained facts which I have indicated. As an illustration of an explainable effect which for a long time troubled the minds of students, the reader may refer to Art. 61—in which I speak of the elastic strength of materials, which to some extent depends upon the loads to which the materials have previously been subjected.

TABLE

Material.	Melting Point. (Fahr.)	Specific Gravity	Weight of One Cubic Foot in pounds.	Expansion by heating from freezing point to boiling point.	Breaking Stress, in per sq.	
					Tensile.	Compressive.
Cast Iron {	2,786° to 2,600°	7·11	444	·0011	{ 30,500 17,500 10,800	130,000 95,000 50,000
Wrought Iron Bars } Wrought Iron Plates, with fibre } Ditto, across fibre Ditto, mean	3,280° to 3,500°	7·7	480	·0012	{ 67,000 57,600 33,500 50,700 46,100 48,400	} 50,000 — — —
Soft Steel, unhardened		—			{ 100,000 80,000 60,000	} —
Ditto, hardened	3,300° to 2,850°	7·8	489	—	120,000	—
Cast Steel, untempered				·0011	{ 150,000 120,000 84,000	} —
Ditto, tempered				·0011	—	—
Copper	2,000°	8·8	556	·0018	33,000	58,000
Brass, Yellow	1,847°	7·8 to 8·4	487 to 524	·0019	17,500	10,500
Gun Metal	1,900°	8·6	536	—	{ 52,000 36,000 23,000	} —
Foundry Metal	—	8·1	505	—	49,000	—
Phosphor Bronze	—	—	—	—	58,000	—
Cast Zinc	758°	7	436	·0029	7,500	—
Lead	600°	11·4	712	·0028	1,900	7,300
Tin	442°	7·4	462	·0028	4,700	—
Wood, Pine	—	·5 to ·7	31 to 44	as glass.	12,000	6,000
Ditto, Oak	—	·7 to 1·0	43 to 62	—	15,000	10,000
Leather	—	—	—	—	4,200	—
Red Brick Fire Brick	—	{ 2 to 2·167	} 125 to 135	{ — ·0005	290 to 300 —	1,100 to 550 1,700
Granite	—	2·7	168	·0009	—	{ 11,000 to 5,500
Cement	—	—	{ 56 (In state of dry powder.)	} —	200 to 600	10,000
Marble	—	2·8	175	·0014	—	5,500
Limestone	—	—	—	—	—	4,250
Sandstone	—	2·3	144	·0018	—	{ 5,500 to 2,200
Plate Glass	—	2·7	169	·00089	9,400	—
Hemp Rope, in ordinary state	—	1·3	—	—	5,600	—
Slate	—	2·8	175	·0014	{ 9,600 to 12,800	} —
Ordinary Mortar	—	—	—	—	50	—
Brickwork	—	1·8	112	—	—	—

III.

pounds inch.	Stress which produces Permanent Set.			Safe Limit of Stress, in pounds per sq. inch.			Modulus of Elasticity, millions of pounds per sq. inch.	Modulus of Rigidity millions of pounds per sq. inch.
Shearing.	Tensile.	Compressive	Shearing.	Tensile.	Compressive	Shearing.		
} 28,500	10,500	21,000	7,900	3,600	10,400	2,700	{ 23 17 14 }	6·3
50,000	24,000	24,000	20,000	10,400	10,400	7,800	29	10·5
—	—	—	—	—	—	—	25	—
—	—	—	—	—	—	—	27	—
—	20,000	20,000	15,000	10,000	10,000	7,800	26	9·5
—	35,000	—	26,500	17,700	17,700	13,000	30	11
—	70,500	—	53,000	—	—	—	30	11
—	80,000	—	64,000	52,000	52,000	38,500	30	11
—	190,000	—	145,000	—	—	—	36	13
—	4,300	3,900	2,900	3,600	3,120	2,300	15	5·6
—	6,950	—	5,200	3,600	—	2,700	9·2	3·4
—	6,200	—	4,150	3,120	—	2,400	9·9	3·7
—	19,700	—	14,500	9,870	—	7,380	14	5·25
—	3,200	—	—	—	—	—	—	—
—	1,500	—	—	—	—	—	·72	·27
650	—	—	—	—	—	—	1·4	·09
2,300	—	—	—	—	—	—	1·5	·08
—	—	—	—	—	—	—	·025	—
—	—	—	—	—	—	—	—	—
—	—	—	—	—	—	—	—	—
} —	—	—	—	—	—	—	—	—
—	—	—	—	—	—	—	—	··
—	—	—	—	—	—	—	—	—
} —	—	—	—	—	—	—	8	—
—	—	—	—	—	—	—	—	—
—	—	—	—	—	—	—	16	—
—	—	—	—	—	·	—	—	—
—	—	—	—	—	—	—	—	—.

CHAPTER VIII.

MATERIALS.

67. A little knowledge is not a dangerous thing if the owner is modest enough to feel that it is only a little. It is often very useful, for instance, to know the most elementary facts of **chemistry**, for these will give you clear ideas as to the changes which occur during the manufacture of metals, the cause of the rusting of metal, the burning of fuel, and many other matters which you would otherwise be unable to comprehend. Again, a little knowledge of **electricity** would enable you to get clear ideas as to the action by which, when two metals touch in a liquid, one of them rapidly corrodes and the other does not, and how it is that oil preserves a polished metal surface. A little knowledge of **heat** will give you clear ideas as to how friction wastes mechanical energy by converting it into heat. It will tell you that when a body is heated it expands uniformly in all its dimensions; wrought iron, ·0001235 of every dimension for one degree Fahrenheit; cast iron, ·00001127; steel, ·00001145; brass, ·00001894; copper, ·00001717; lead, ·00002818; glass, ·00000861; and Platinum, ·00000884. It will tell you that when a gas is heated 490 degrees from the temperature of freezing water at constant pressure, it expands to twice its volume or cubic content, and that liquids expand very much less than gases and more than solids. It will also give you clear ideas about melting and boiling, about the way in which heat is measured as a form of energy, and the properties of steam which enable it to be used in the steam-engine. It will also tell you about the giving up of heat from one body to another by conduction and radiation, things which enter into every process going on in the workshop, and of which you can only have vague and incorrect ideas unless you spend a month or two in experimenting. I am sorry I cannot give you this clearness of ideas by anything which I can write; I know no other way of obtaining it than through your own handling of some simple apparatus such as is usually kept unfortunately for the mere illustration of lectures.

I mean in this chapter to give a rough account of the various materials used in construction.

68. Stone.—The rocks which have once been melted, and have cooled slowly, are usually hard, compact, strong, and durable. They are most easily worked when regard is paid to the fact that they naturally divide up into certain regular shapes. They are all more or less crystalline in texture. *Stratified* rocks are those which have been deposited at the bottom of a sea or river; they are often easily divided in a direction parallel to the layers of which they are built up, but sometimes there are lines of easy cleavage in other directions. These rocks vary very much in appearance, according to the method of their formation, and to the heat and pressure to which they have been subjected, sometimes being very crystalline, strong and durable, like *marble; slaty* rocks may be hard and durable, or soft and perishable; *sandstones* are hardened sand of different degrees of compactness, porosity, strength, and durability; there are *limestones* whose particles seem to form one continuous mass, and which, when they have been subjected to great heat and pressure, become *marbles;* there are also limestones, which are composed of distinct grains cemented together, and which may vary very much in compactness, strength, and durability; besides these there are *conglomerates,* in which fragments of older rocks are imbedded. A little knowledge of **geology** is necessary in order to understand the properties of rocks. **Stones are preserved** by coating them with some material such as coal-tar, various kinds of oil and paint, and soluble glass, which fills their pores and prevents the entrance of moisture. An **artificial stone**, which can be made in blocks of any required size and shape, is obtained by turning out of moulds and afterwards saturating with a solution of chloride of calcium, a mixture of clean sharp sand and silicate of soda. The chloride of calcium and silicate of soda produce silicate of lime which cements the sand together and thus gradually consolidates the whole mass.

69. **Bricks.**—Bricks are made of tempered clay, moulded, dried gently, then raised to and kept at a white heat in a kiln for some days, and cooled gradually. Bricks should have plane parallel surfaces and sharp right-angled edges, should give a clear ringing sound when struck, should be compact, uniform, and somewhat glassy when broken, free from cracks, and able to absorb not more than one-fifteenth of their weight of water. They ought to require at least half a ton per square inch to crush them.

70. **Limestone**, when burnt in kilns, gives off carbonic acid. If pure it forms *quick-lime*, which combines readily with water, becoming larger in volume. Mixed with clean sand this forms **mortar**, which, in the course of time, hardens by losing its water and combining with carbonic acid from the air. If the burnt limestone were not pure, but contained certain kinds of clayey materials, iron, &c., it would not combine with much water, but when ground up fine, water enables its particles to combine chemically with one another with greater or less rapidity, depending on its composition. Such **cement** first *sets*, acquiring a large degree of firmness, and then more slowly becomes as hard as many limestones. When these natural hydraulic limestones are not available, nearly pure limestone may be mixed with a proper proportion of blue clay to produce, when ground and mixed in plenty of water, then drained and dried, then burnt and ground up again, an artificial cement, which is equal, if not superior, to the natural cement. *Sand* in mortar

saves expense, and prevents the cracking of the mortar in drying, but in too great a proportion it weakens the mortar. Two measures of sand to one of slaked lime in paste is the average allowance, but every person who uses mortar ought to test a particular lime to see how much sand it will bear to have mixed with it. **Concrete** is a mixture of gravel or broken stones and hydraulic lime, the stones and gravel having about six times the volume of the lime.

71. Earth.—It is usual to consider that the pressure of earth against a wall, A B (Fig. 34), is due to the tendency of a wedge-shaped mass of earth to slide downwards. We may suppose that A B C, or A B D, or A B E, is the sliding wedge, and we choose for our calculation that one which presses most against the wall. It is the weight of the wedge of earth which urges it downwards; friction at its

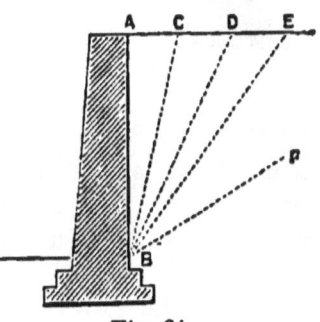

Fig. 34.

face B C, B D or B E tends to support it, as well as friction against the surface A B, where it presses on the wall. This friction is usually calculated from knowing B F, the *natural slope* taken by the earth when not prevented from sliding. It is obvious that if the earth is very soft, or if much water gets between the earth and the wall, the pressure becomes like that of water. It cannot be said that experience has proved the untruth of this old theory; experience has shown that it is somewhat difficult to find what is the natural slope of the earth immediately behind a wall, and what is the friction between the wall and the earth. Rankine, neglecting the friction against the wall, obtains from such a common-sense view as I have given, the following rule, which has been found to work fairly well in practice. Draw an

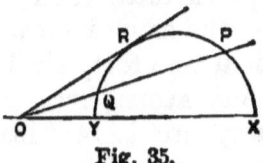

Fig. 35.

angle X O R (Fig. 35) to represent the natural slope of the earth. Describe Y R X a semicircle touching O R. Now if A B (Fig. 36) is the vertical face of a wall sustaining a

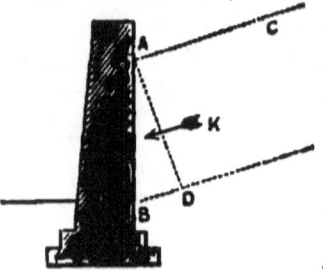

Fig. 36.

bank of this earth whose slope is A C, make the angle X O P equal to the inclination of A C to the horizon. Find B D so that P O : O Q :: A B : B D.

Then A B D is a wedge of earth whose *weight* represents the *total pressure* acting on A B. The pressures act in directions parallel to A C, and the resultant force, representing the total pressure, acts a third of the way up from B to A. You must remember that this is a mere rule giving the result of a calculation, and that the wedge A B D is an imaginary thing used to help the memory.

72. Water.—The pressure of still water is at right angles to any surface, and does not depend on the slope of the surface. It is greater at greater depths. If the pressure per square foot at any place is known, we can calculate the additional pressure at any lower level, for it is the weight of a vertical column of water one square foot in cross section reaching from the one level to the other. The pressure at all places on the same level is the same. Suppose that when water fills a vessel from which it cannot escape, we push in a piston or plunger until the pressure on the plunger is increased by say ten pounds per square inch, then at every place in the vessel there will be the same increase of pressure. Water is compressed about one forty-sixth-millionth of its cubic content for one atmosphere of pressure. (The pressure of one atmosphere is 14·7 lbs. per square inch, or 2,117 lbs. per square foot.) The total pressure of water on any surface is obtained by regarding the pressure on each little portion of its area as a force, and finding the resultant of all the forces. On any plane surface submerged in a pond, *the total pressure is found to be the weight of a column of water whose cross section is the area, and whose length is equal to the vertical depth of the centre of gravity of the area below still water level.* If water is not still, but has a steady motion of any kind, let us consider the path taken by any particle. Suppose that it goes more quickly at one place than another, then we shall find that its gain of kinetic energy is accompanied by a lowering of level or else by a lessening of pressure. If it is not getting lower in level then it must be exerting less pressure. In a horizontal pipe where the section is smaller the velocity must be greater, and here the pressure must be less.* No matter how quickly water may move in a pipe, the pressure can never become equal to that of a vacuum, because the water will give off vapour and completely alter the conditions of the case. Remember that the law given in the note supposes that there is no friction. The frictional loss of energy experienced by a particle of water moving in pumps and pipes is found by experiment to be nearly proportional to its kinetic energy.† Hence in hydraulic presses, and

---

* If a little volume of water (one cubic foot we take for simplicity), whose weight is w pounds, is h feet above some datum level, if the pressure upon it is p pounds per square foot, and its velocity is v feet per second, then hw is the potential energy, due to its merely being above the datum level. It has also, in virtue of the steadiness of the motion, pressure or potential energy, which is represented by p foot-pounds, and its kinetic energy is $\frac{1}{2}\frac{w}{g}v^2$. Its total energy is then—

$$hw + p + \tfrac{1}{2}\frac{w}{g}v^2,$$

and however its position, pressure, or velocity may change during its motion, the sum of these three terms remains the same, so that if two are given the third may be calculated. The student may object to this by saying that pressure cannot be regarded as a form of energy; however, it is certain that in steady motion pressure enters into the expression for the total energy, and this is due to the fact that in nearly still water the pressure represents the work which all the rest of the water will do upon a particle should it rise slowly to a higher level.

† The *force* of friction in fluids is proportional to the velocity, when the velocity is small; it is proportional to the square of the

in other machines where there is only a slow motion of the water, the loss through friction is much more negligible than it is in turbines and pumps. Thus, in a reciprocating pump, as the flow of the water is stopped in the barrel and valve-chest every stroke, its kinetic energy is all wasted, and hence it is advisable to make this flow as slow as possible. By the use of air vessels we can prevent the flow of water being suddenly stopped, and thus prevent the total loss of the kinetic energy. At any particular kind of bend in a pipe the energy lost is a certain fraction of the kinetic energy, and this fraction is found by experiment.

73. *Example.*—Water flows from an orifice in a vessel into the atmosphere. The free water surface is twelve feet above the orifice. What is the velocity of a particle of the issuing water which is in contact with the atmosphere? (The particles in the interior of the jet may not be at the pressure of the atmosphere.) Now, when this particle was motionless at the surface of the water in the vessel, its pressure was that of the atmosphere; call it zero. Pressure energy, then, is zero at beginning and end. Loss of potential energy is the weight of the particle multiplied by the difference of level, and this has all been converted into kinetic energy. If the weight of a particle is 1 lb., it has 12×1 or 12 foot-pounds of potential energy changed into kinetic energy, but its kinetic energy is $\frac{1}{64 \cdot 4} \times$ square of its velocity in feet per second; hence the square of its velocity is $64 \cdot 4 \times 12$, or $772 \cdot 8$, or the velocity is $27 \cdot 8$ feet per second. You will, in fact, find that the velocity of the particle is the same as if it had fallen

velocity in the case of ordinary steamers, and becomes proportional to a higher power of the velocity in very quick moving vessels. Now the *energy wasted per second* in overcoming friction is equal to the *force of friction multiplied by the velocity per second*. Hence in water pipes, when the velocity is not great, the energy lost is proportional to the square of the velocity; in ordinary ships it is proportional to the cube of the velocity.

freely from the height of twelve feet. If, instead of flowing into the atmosphere, the water flowed into a place where the pressure is greater than that of the atmosphere, the velocity would have been less. If you can find a place in the issuing jet at every point of which the water flows at right angles to the cross section of the jet, and this seems to be the case at the most contracted part of the jet just outside a circular orifice, then *the area of this cross section in square feet multiplied by the velocity we have calculated in feet per second gives the quantity of water in cubic feet per second.* In the same way *the quantity of water flowing through a pipe is the cross sectional area in square feet multiplied by the velocity.*

74. **Timber.**—A tree is made up of a great number of little tubes and cells arranged roughly in concentric circles. The process of seasoning consists in uniformly drying the timber. As each little portion dries, it contracts, and becomes more rigid, and it contracts much more readily in the direction of the circular arrangement of the tubes than it does towards the centre of the tree, and least easily in a direction along the tree. It is obvious, then, that if the tree is dried whole, there will be a tendency to splitting radially. If the tree is cut up before drying we can tell the way in which the planks will warp if we remember the above facts.

Firwoods are easily wrought, and possess straightness in fibre and great resistance to direct pull and transverse load, and are largely used because of their cheapness. They differ greatly in strength, but their weak point is their inability to resist shearing. The best of these is the *red pine* or *Memel* timber from Russia, which can be had in large scantlings, and thus used without trussing. The *white fir* or *Norway spruce* is suitable for planking and light framing, and is imported from Christiania in "deals," "battens," and "planks." *Larch* is a very strong timber, hard to work, and has a tendency to warp in drying, and is therefore not suitable for framing, but is largely used for railway-sleepers and fences, because of its durability when exposed to the weather. *Cedar* lasts long in roofs, but is deficient in strength.

The English *Oak* is the strongest and most durable of all woods grown in temperate climates, but is very slow-growing and expensive. Its great durability when exposed to the weather seems to be due to the presence of gallic acid, which, however, in any wood corrodes iron fastenings; trenails or wooden spikes should be used instead. *Teak*, which is grown in the East, is the finest of all woods for the engineer. It is very uniform and compact in texture, and contains an oily matter which contributes greatly to its durability. It is used specially in ship-building and railway carriages. *Mahogany* is unsuitable for exposure to the weather, but it has a fine appearance and is not likely to warp much in drying. It is chiefly used for furniture and ornamental purposes, and to some extent in pattern-making. *Ash* is noted for its toughness and flexibility, and a capability of resisting sudden stresses of all kinds, which make it specially adapted for handles of tools and shafts of carriages. It is very durable when kept dry. It is not obtainable in large scantlings, and is sometimes very difficult to work. *Elm* is valuable for its durability when constantly wet, which makes it useful for piles or foundations under water. It is noted for its toughness, though inferior to oak in this respect, as also in its strength and stiffness. It is very liable to warp. *Beech* is smooth and close in its grain. It is nearly as strong as oak, but is durable only when kept either very dry or constantly wet. It is very tough, but not so stiff as oak. (See also Table VI.)

The **best time for felling timber** is when the tree has reached its maturity, and in autumn when the sap is not circulating. We want to have as little sap in the timber as possible, and in order to harden the sapwood, some foresters are of opinion that the bark should be taken off in the spring before felling. After timber is felled, it is well to square it by taking off the outer slabs.

Timber is, for the most part, dried by putting it into hot-air chambers, from one to ten weeks according to the thickness. Even when kept quite dry, ventilation is necessary to prevent dry rot. The circumstances least favourable to the durability of timber are alternate wetting and drying, as in the case of timber between high and low water mark, whereas good seasoning and ventilation are most favourable conditions. The most effective means adopted for **preserving timber** is by saturating it with a black oily liquid called *creosote*. The timber is placed in an air-tight vessel, and the air and moisture extracted from its pores as far as possible. The warm creosote is then forced into these pores at a pressure of 170 lbs. per square inch. In this way timber may be made to absorb from $\frac{1}{10}$th to $\frac{1}{12}$th of its weight of creosote.

75. Glass.—Glass is a combined silicate of potassium or sodium, or both, with silicates of calcium, aluminium, iron, lead, and other chemical substances. Certain mixtures of flint and chemicals are melted in crucibles, formed when hot into the required shapes, and cooled as slowly as possible. The more slowly and more uniformly the cooling is effected, the more likely is it that the glass shall be without internal strains. When glass is suddenly cooled, as when a melted drop falls into water, the outside is suddenly contracted, becomes hard and brittle, and there are such internal strains that if the tapering part be broken or scratched at the point, the whole drop crumbles into a state of dust. A blow or scratch on the thick part produces no such effect. Heating and gradual cooling destroys this property. Many peculiarities in the behaviour of metals when heated and cooled seem to be caricatured in glass, possibly because they are due to the fact that all the portions of matter which are about to form one crystal must be at the same temperature, and when the substance is a bad conductor of heat there is great variation in temperature. Pure metals are good conductors, but the admixture of small quantities of carbon and of gases hurts their conductivity. *Toughened glass* is the name wrongly given to the hardened glass produced by plunging glass, in a nearly melting state, into a rather hot oily bath. This glass is somewhat in the condition of the glass in a Rupert's drop. It is so hard that it is difficult to cut it with a diamond, but if the diamond cuts too deep the whole mass breaks up into little pieces. Objects made of it may be thrown violently on the floor without breaking.

76. Cast Iron.—Certain chemical changes occur when the ores of iron are smelted; the iron ceases almost entirely to be in chemical combination with other substances, and impurities almost disappear, excepting *carbon*, which is mainly derived from the fuel. In the cupola of the foundry a greater purification is effected, and it is found that the composition of a casting is from 97 to 95 per cent. of iron, with 3 to 5 per cent. of carbon, although traces of other substances are to be found. About $2\frac{1}{2}$ cwts. of good coke are usually required to melt each ton of iron in a cupola. When the *carbon is all chemically combined with the iron, the cast iron is white (specular iron) and is very hard and brittle. When only a little of the carbon is chemically combined, and most of its particles crystallize separately, the cast iron is grey in colour.* Using the common names for the different varieties, No. 1 is darkest in colour, and from No. 4 to No. 1 there is a gradual darkening in colour. Nos. 1, 2, 3, and 4 are commonly used in the foundry, mixtures being made of them in various proportions according to circumstances. A greater proportion of No. 3 or 4 gives greater strength, whereas a greater proportion of No. 1 gives greater fluidity, and a better power of expanding at the moment when the metal solidifies, so that the sharp corners of the mould are better filled. Higher numbers than 4, as 8, 7, 6, and 5, the white varieties, are seldom used in the foundry, but they may be converted into grey varieties by slow cooling. To soften a hard casting it is heated in a mixture of bone ash and coal dust or sand, and allowed to cool there slowly.

77. Patterns of objects are usually made in yellow pine, about one-eighth of an inch per foot in every direction larger than the object is to be, because the iron object contracts to this extent in cooling. *Prints* are excrescences made on the patterns to show in the mould where certain *cores* are to be placed. These cores are made of loam or core sand in core-boxes, which the pattern-maker supplies; they represent the spaces in the object where the melted metal is not to flow. You must see for yourself in a foundry what are the usual methods of preparing a mould; how the pattern is made so as to draw out easily; how the moulder arranges his *vents* to let gases escape; how he places his *gates* to let the metal run into the mould with just enough rapidity, and yet without hurt to the mould. You must also see for yourself, taking sketches in your notebook and making a drawing of the *cupola*, how the pig iron is melted and poured into the moulds; how the moulder stands moving an iron rod up and down in one of the *gates*, producing just so much circulation and eddying motion in the melted iron, as is likely to remove bubbles of gas which may otherwise be unable to escape from the sides and corners of the mould; how in some castings he exposes to the air certain parts which would otherwise cool too slowly for the rest of the object; how next morning he *screens* his sand and wets it. You ought to observe the appearance of the castings before and after they are cleaned up next morning.

78. The Cooling of Castings.—The most important matter in connection with moulding is that there shall be the same amount of contraction at the same time in every portion of the mass of metal as it cools; otherwise, when finished, there may be *internal strains* which very much weaken the object, and often produce fracture. In designing the shape of an object which is to be cast, care is taken that when a thin

G

portion joins a thick one, it shall do so by getting gradually thicker, and not by an abrupt change of size. The thin piece exposes more surface, and cooling is effected through the surface. The thin rim of a pulley cools sooner than the arms, and becomes rigid sooner: when the arms cool they contract so much as sometimes to produce fracture near the junction. In a thick cylindric object the outer portion becomes rigid first; now, when the inner portion contracts, it tends to make the outer portion contract too much, and the outer portion prevents the inner from contracting as much as it ought to, so that the outer portion retains a compressive strain, and the inner a tensile strain. When a hollow cylinder is cast, and is required to withstand a great bursting pressure, that is, all the metal is required to withstand tensile stresses, it is usual to cool it from the inside by means of a *metal core*, in which cold water circulates. The inside now becomes rigid sooner, the outer portions as they solidify contract, and tend to make the inner portion contract more than it naturally would, and there is a permanent state of compressive strain in the object which materially helps it to resist a bursting pressure. This inequality of contraction and production of internal strains in objects cause them to vary in their total bulk as compared with that of their patterns, but it is probable that some of this variation is due to the fact that the contraction of grey cast iron is only one per cent. of its linear dimensions, whereas white cast iron contracts two to two and a half per cent. The fractional difference between size of pattern and the finished object varies from one-twenty-fifth of an inch per foot in small thin objects, to one-eighth of an inch per foot in heavy pipe castings and girders. As there is always great inequality in the rate of cooling of a casting near a sharp corner, internal strains may be expected here, and also an inequality in the nature of the cast iron, since the grey variety gets whiter the more rapidly it is cooled; now, in nearly all bodies a re-entrant corner is a place of weakness (see Art. 95), and is specially to be guarded against in castings. Crystals of cast iron and other metals group themselves along lines of flow of heat. When a plate or wire of iron or steel is rolled or pulled, the crystals become more longitudinal, and the wire or plate becomes stronger, whereas annealing allows the crystals to arrange themselves laterally, and the material is weakened. Castings which have been rapidly cooled by being cast in an iron mould (painted on its inside with loam) are *white*, and very hard in those parts which lie nearest the mould, whereas they are grey and strong inside. These are called *chilled castings*. When a casting is put in a box, surrounded with oxide of iron, and kept at a high temperature for a length of time, its surface, to a depth dependent on the time, loses its carbon and becomes pure or wrought iron, which is much tougher than cast iron. The teeth of wheels are sometimes heated in this way. Such are *malleable castings*. Melted cast iron possesses the property of dissolving pieces of wrought iron, and is then said to be *toughened cast iron*.

79. Wrought Iron.—Cast iron is exposed to the air in a melted state for a long time, and the carbon is burnt out of it. The pig-iron really undergoes two processes, one called *refining*, the other *puddling*. It is then hammered and rolled when hot into bars of various shapes. The quality of wrought iron bars as bought in the market varies greatly. We have *common* iron, used for rails, ships, and bridges; *best, double best*, and *treble best* Staffordshire iron, used for boilers and forgings generally; Lowmoor, Bowling, and other good irons for the most difficult forgings; and lastly, *charcoal* iron, which is nearly pure. Up to the temperatures of ordinary boilers, the tensile strength of iron is not much diminished by heating, but at a red heat it is very much less than in the cold state. By rolling and hammering when hot, iron gets a fibrous texture, and becomes more tenacious. By hammering when cold, or by long continued strains of a vibratory kind, it is thought that wrought iron changes its fibrous and tough for a crystallised and more brittle condition. This brittle condition may be removed by heating and slowly cooling (annealing). Iron wire is stronger the thinner it is. Bar iron is generally stronger than angle or T-iron, and this again than plate iron. The toughness of an iron bar is best shown by the contraction it undergoes before it breaks. The section of a very tough bar may contract as much as forty-five per cent. in area. *Case hardening* of a wrought iron object is effected by heating it in a box with bone-dust and horn shavings. The iron absorbs carbon, and is partially converted into steel.

80. **Steel.**—Steel contains less carbon and impurities than cast iron, and thus lies intermediate between cast iron and wrought iron. It is produced by giving carbon to wrought iron, keeping the iron heated for some days in contact with powdered charcoal, and then hammering it whilst hot till it is homogeneous, or else casting it when melted into ingots. Steel is also produced by taking only a portion of the carbon from very pure varieties of cast iron by a puddling process such as is employed in the production of wrought iron, or by the Bessemer process. In the Bessemer process, air is forced into the melted cast iron for a time, and very pure white cast iron is then added to help in removing bubbles of gas. I have already told you about the tempering of steel. (Art. 65.) It is more fusible than wrought iron, and some success has been met with in the production of steel castings in spite of the fact that they are apt to contain cavities. The strength of steel is greater than that of any other material, and is greater as it contains more carbon. The properties of steel depend so much on so many seemingly small things, small impurities, a little too much heating or variation in the rate of cooling at different places, that great care must be taken in working it. By the Bessemer and Siemens processes great quantities of steel are produced cheaply, contain-

ing small percentages of carbon. This steel is largely coming into use for locomotive rails, bridges, and ships, instead of wrought iron.

81. Copper is noted for its malleability and ductility when both hot and cold, so that it is readily hammered into any shape, rolled into plates, and drawn into wires. When cast it usually contains many cavities, but when pure it may be worked up by hammering into a state of great strength and toughness, whereas slight traces of carbon, sulphur, and other impurities necessitate its being refined before it loses its brittleness. The brittleness produced by hammering when cold is very different, as it is removable by annealing. Copper is an expensive metal, and is only used now for pipes which require to be bent cold, for bolts and plates in places where iron would be more readily corroded, and for electrical purposes. Its tensile strength is more reduced by heating than that of iron.

82. Brass consists of about two parts by weight of copper to one of zinc. It is used chiefly on account of its fine appearance and the ease with which it can be worked. A little lead added in melting makes it much softer. **Muntz metal** contains more zinc than ordinary brass. **Bronze** and **Gun-metal** are alloys of copper and tin in varying proportions, more tin giving greater hardness. Five of copper to one of tin is the hardest alloy used by the engineer. A slight addition of zinc increases the malleability. A great many experiments have been made on bronze. Its strength depends very much upon the care taken in mixing the metals. It makes good castings, which are usually made in cast-iron moulds. Hard bronze is much used for the bearings of shafts. There are also various *soft alloys* of copper with lead, zinc, tin, and antimony, which are used for this purpose. **Phosphor bronze** is an alloy of copper and tin, to which some phosphorus has been added. It bears re-melting (unlike gun-metal), and its properties may be varied at will. It may be either strong and hard, or weaker but very tough.

CHAPTER IX.

SHEAR AND TWIST.

83. Let C D (Fig. 37) be the top of a firm table, F H a long prism of india-rubber glued to the table, A B a flat piece of wood glued along the upper side of the india-rubber. We try in this way to apply a horizontal force to the whole upper surface of the india-rubber, so that if, for instance, the pull in the cord is 20 lbs., and the upper surface of the

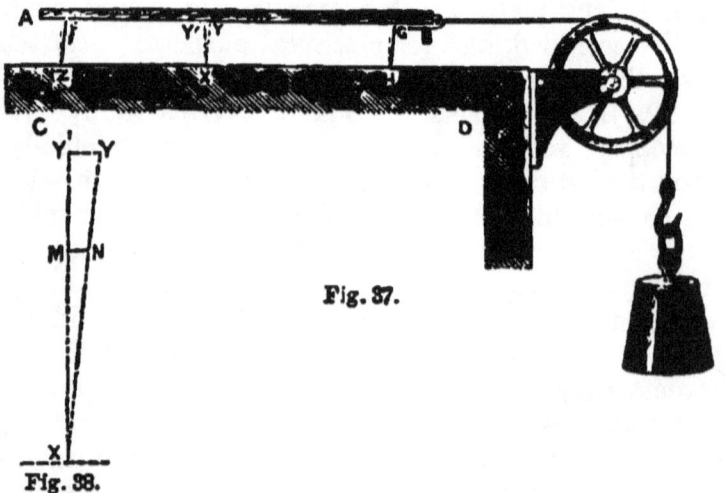

Fig. 37.

Fig. 38.

india-rubber is 10 square inches in area, there will be a force of 2 lbs. per square inch acting at every part of the surface, and this force will be transmitted through the india-rubber to the table. When the length of the prism is great compared with Z F, we may suppose that the bending in it is very small, and in this case we say that the india-rubber is being subjected to a pure *shear strain*, and the force per square inch acting on its surface is also acting from each horizontal layer to the next and is called the *shear stress.* If you had drawn vertical lines like Y' X before the cord was pulled, you would now find them sloping like Y X. Thus, making a magnified drawing of Y X in Fig. 38, the point Y' has gone to Y, and any point like M has gone to N. Points touching the table cannot move, but the farther a point is away from this fixed part the further it can move. Now suppose that Y' Y is 0·01 inch, and we know that X Y' is 2 inches, what is the amount of motion of M if M X is 1·7 inch? Evidently Y' Y is greater than M N just in the proportion of Y' X to M X, or 2 to 1·7, hence M N is 0·0085 inch. Thus the motion of any point is simply proportional to its distance above the fixed plane, and if we know the amount of motion at, say a distance of one inch, we can calculate what it must be anywhere else. *The amount of motion at one inch above the fixed plane is called the* **shear strain**. In this case we have supposed the force on F G to be 2 lbs. per square inch. This is said to be the amount of the **shear stress**, and it produces or is produced by a *shear strain* whose amount is ·005 inch per inch. If the shear stress were 4 lbs. per square inch, you would find the strain to be ·01, if the stress were 8 lbs. per square inch the strain would be ·02. In fact, we find experimentally that the stress and strain are proportional to one another. Thus if, instead of india-rubber, we had a block of tempered steel, we should find that the force in pounds per square inch is equal to 13,000,000 times the strain. This number is called the *modulus of rigidity* for steel; it is given in Table III.

84. *Example.*—A beam of steel has one end fixed, and at the other is a weight of 20 tons. The cross section of the beam is 2 square inches in area, and the length of the beam is 5 inches. Besides the deflection of this beam due to bending, there is a certain **deflection due to shearing**; how much is it? Answer: the shear stress is 10 tons, or 22,400 lbs. per square inch. This produces a shear strain of 22,400 ÷ 13,000,000, or ·00172. This is the amount of yielding at 1 inch from the fixed end, and at 5 inches the yielding must be 5 × ·00172, or ·0086 inch.

85. The shear stress which will **produce rupture** is not well known for any substance except cast and wrought iron, but the shear stress which will produce permanent set is fairly well known, and we are also agreed as to the ordinary working shear stress of materials. For wrought iron it is usually regarded as less than the working tensile stress; but in a single-riveted lap joint in boiler-plates, as the holes are usually punched (and this weakens the metal), and as rivet iron is usually of a better quality than plate, the cross section of the iron which is left, which is resisting pull, is made to have the same area as the cross sections of all the rivets, which, of course, resist shearing. Besides breaking by either a tensile or a shear stress, a riveted joint may give way by the rivet crushing or being crushed by the side of its hole. Again, in many riveted joints, when the rivets are long, as they tend to contract in cooling and are prevented by the plates, so much tension may remain permanently in them that they are greatly weakened. In bolts there is usually a want of perfectly uniform distribution of the shear stress, and they are made larger than rivets in the same positions.

86. In the **punching** of rivet-holes it is a shearing force which acts on the material; *the area of the curved side of the hole, multiplied by the breaking shear stress of the material per square inch, represents the force with which the punch must be pressed down on the plate.* The punch must be able to resist this force as a compressive stress on its own material. Experiments made on punching machines show that about 24 tons per square inch is the average shearing force required. This pressure has to be exerted through a very short distance indeed, for as soon as fracture occurs the punch has to overcome no more resistance to shear. In shearing machines, if the entire edges of the shears coincided

with the plate, as soon as they touched anywhere there would be the same sort of effect produced; but by inclining the edges the shearing action does not occur instantaneously at every place, and the rupture being more gradual than in punching, the shearing resistance is usually from 10 to 30 per cent. less. It is very probable that the power lost in punching and shearing machines is wasted rather in the friction of the heavy parts of the mechanism than in the almost instantaneous effort of cutting the material. The effort required seems rather that of an impact (see Chap. XIX.) than of the more gradual action to be found in most existing machines. The only excuse for using such uneconomical machines as hydraulic bears and shears is that, although they are uneconomical, they may be worked by hand. In the fly presses used for hand-punching, and used largely in coining, the idea of an impact is already in use; it will come much more into use in large machines when engineers become better acquainted with the distinction between force and energy.

87. However long we may make our block of india-rubber in Fig. 37, we shall still have some bending in it—that is, the stress will not be uniformly distributed over each horizontal layer. To prevent this bending effect, and

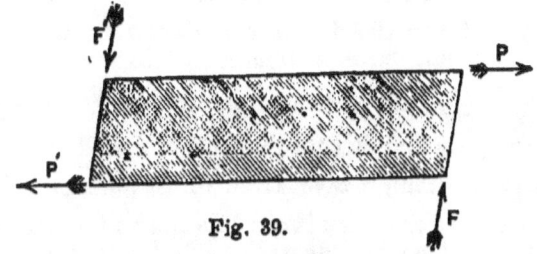

Fig. 39.

to produce a really pure shear strain, we ought to have force distributed over the ends F Z and G H of the same amount per square inch as we have now acting over F G and Z H. These are shown in Fig. 39, where P is the pull in the cord of Fig. 37, P' is the equal and opposite force exerted by the table on the glued under-side of the india-rubber, and F and F' are equal and opposite forces distributed over the ends, such that the couple F F′ is able to balance the couple P P′. There can now be no bending moment at any place. As F multiplied by the length of the prism is the moment of the couple F F′, and is equal to P multiplied by the vertical dimension, we see that P distributed over the horizontal surface is the same stress per square inch as F distributed over the ends. From such a material then, if we cut a cubical block, A, Fig. 40, its horizontal faces Y y and x x are acted upon by equal and opposite tangential forces, and its faces Y x, y x are acted upon by forces of exactly the same amount. The faces parallel to the paper have no forces acting on them. This will give you the best idea of **pure shear stress**.

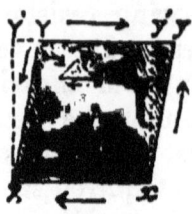
Fig. 40.

The material in Fig. 37 near the ends of the block does not get a pure shear, but if the block is very long, then at the middle there is a nearly pure shear acting.

In Fig. 40 the cube x Y′ y′ x has become x Y y x. Suppose the side of this cube to be 1 inch, then Y′ Y is the shear strain, which I shall call s. The tangential force distributed over Y y is p lb., let us say. Then if we denote by the letter N the **modulus of the rigidity** of the material,

$$p = N\ s.$$

88. Nature of Shear Strain.—Now when y′ moves to y, the diagonal x y′ becomes extended to x y. Its original length was $\sqrt{2}$ (the diagonal of a square whose side is 1) and its new length is $\sqrt{2} + \frac{s}{\sqrt{2}}$, as we see very easily. Hence the diagonal x y′, and all lines parallel to this diagonal, have a tensile strain, whose amount is $\frac{\text{elongation}}{\text{original length}}$ or $\frac{s}{\sqrt{2}} \div \sqrt{2}$, and this is $\frac{s}{2}$. Again, in the same way we find that the diagonal Y′ x, and all lines parallel to it, have a compressive strain whose amount is $\frac{s}{2}$. T h u s it h a s b e c o m e q u i t e c l e a r t h a t a p u r e s h e a r strain simply consists of a compressive strain in one direction, accompanied by a tensile strain in the perpendicular direction, these strains being each half the shear strain. Now when we have a compressive or tensile strain we know that there is compressive or tensile stress which produces it; let us find how much this is. Consider a small right-angled prism of material, shown in Fig. 40, of which M F N, Fig. 41, is the magnified cross

NATURE OF SHEAR STRAIN.

section. Make M F, say 1 inch, M K the same, and let the length of the prism at right angles to the paper be also 1 inch. Neglecting its own weight, this prism is kept at rest by the matter outside it acting on its three faces. Face M F is pushed by a normal force of p' lb. per square inch, and as its area is just 1 square inch, the total push is p' lb. Similarly the face M K is pulled by a normal force of p' lb. And also the face F K is acted on by tangential forces of p lb. per square inch, and as its area is $\sqrt{2}$ square inch, the total amount of shearing force acting on F K is $p\sqrt{2}$ lb. Now

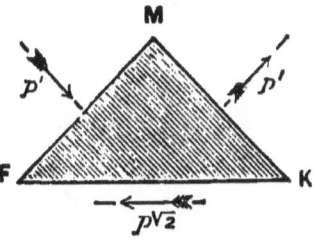

Fig. 41.

when three forces keep a body in equilibrium, and two of them are at right angles, the sum of their squares is equal to the square of the third force (this is easily seen if we draw the triangle of forces), hence the square of $p\sqrt{2}$, which is $2p^2$, is equal to $p'^2 + p'^2$ or $2p'^2$. Hence $p=p'$, and we have proved that **the compressive and tensile stresses which occur in pure shear strain are numerically equal to what we called the shear stress.**

89. One other proof, and I shall leave these interesting theoretical considerations. Suppose we cut a cube, A B C D, Fig. 42, of one inch side from a material subjected to pure shear strain, and let the faces of the cube parallel to the paper have, as before, no stress upon them, the other faces being at right angles to the directions of compression and extension. Shear occurs parallel to the face A C; let us consider the motion of the point D relative to A C; in fact, regard A C as fixed. Under the sole action of the pushes on A D and B C we know that the side D C shortens by the small amount $p a$ (see Art. 54). Let us set this off from D to M. But when this occurs the side A D lengthens by the amount $p b$; set this off from M to D'. Hence the pushing forces on A D and B C cause D to move to D'. Again, the pulling forces on D C and A B further lengthen A D by the distance $p a$, which we set off from D' to L, and shorten D C by the distance $p b$, which we set off from L to D''. Hence the motion of D due to the pulls and pushes acting together is D D'', and we see that this is

$$(\text{D M} + \text{L D}'')\sqrt{2} \text{ or } (a+b)p\sqrt{2}.$$

But s the amount of shear is D D'' $\div$ D O, and as D O $= \dfrac{1}{\sqrt{2}}$ inch and A D is one inch, we have

$$s = (a+b)\, p\, \sqrt{2} \div \frac{1}{\sqrt{2}} \text{ or } 2p\,(a+b)$$

That is, *shear strain* = *shear stress multiplied by* $2\,(a+b)$.

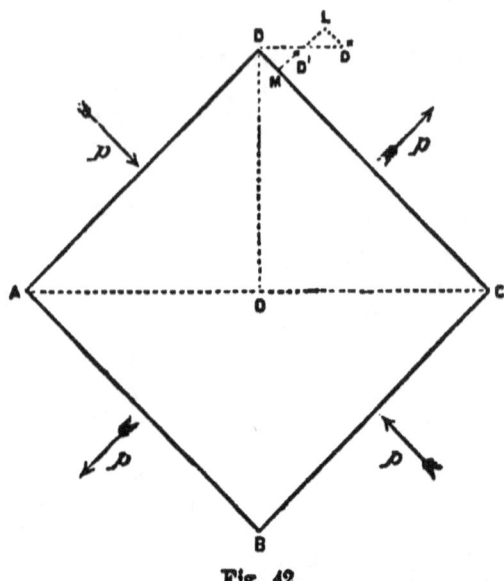

Fig. 42.

So that *the reciprocal of* $2\,(a+b)$ *is what we called* N, the **modulus of rigidity** of the material.

90. General Results.—Referring back to Arts. 54 and 55, you will see that we have

Modulus of rigidity . . . $N = \dfrac{1}{2(a+b)}$

Modulus of elasticity of bulk . $K = \dfrac{1}{3(a-2b)}$

Young's modulus of elasticity . $E = \dfrac{1}{a}$

and you will also see that if we know two of these for any material we can find the third.

Some French mathematicians have thought that the ratio of a to b, and therefore the ratios of N, K and E to one another, are constant for isotropic substances; a being always four times b. Experiment has shown that this is not the case, the ratio of a to b being 3 to 2·5 in glass or brass, 3·3 in iron, 4·4 to 2·2 in copper, and in other substances varying from these values very much indeed.

Just as Young's modulus is seldom found from experiments on the extension of wires, but rather from the bending of beams; so the modulus of rigidity is seldom found from experiments like that of Fig. 37, but rather from experiments on the torsion of rods.

91. Twisting.—In Fig. 43, A B represents a wire held firmly at A. At B there is a pulley fixed firmly to the wire, and this pulley is acted upon by two cords, which tend to turn it without moving its centre sideways. In fact, they act on the pulley with a turning moment merely. But the pulley can only turn by giving a twist to the wire, and the amount of motion it gets tells us how much the twist is. A little pointer fastened at C moves over a cardboard dial, and tells us accurately how much twist is given to the wire. The angle turned through by the pointer is called the *angle of twist* at C. If we had a pointer at each of the places G, H, and C, and if A, G, H, and C were one foot apart from one another, we should find that the angles of twist at G, H, and C are as 1 : 2 : 3; in fact, *the angle of twist is proportional to the length of wire twisted.*

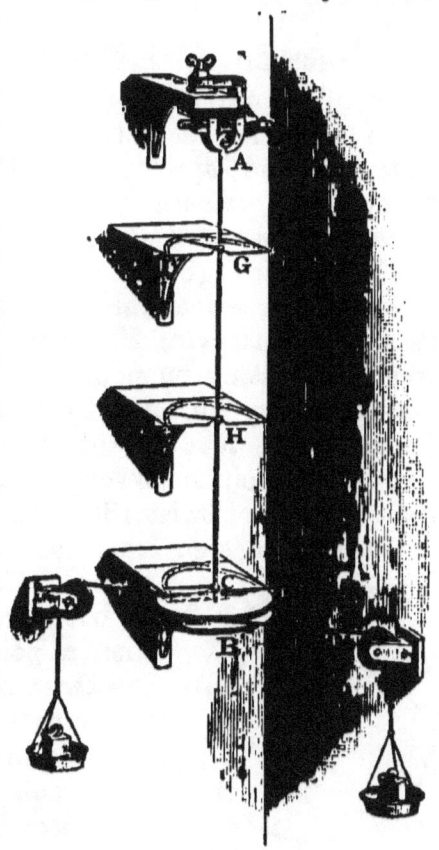

Fig. 43.

You will find that if a twisting moment of 10 pound-feet produces a twist of 4°, then a twisting moment of 20 pound-feet produces a twist of 8°, and, in fact, *the twist is proportional to the twisting moment* which is applied. You will also find that if you try different sizes of wire of the same material, say wires whose diameters are in the proportion of 1, 2, 3, &c., and to each of them you apply the same twisting moment, the amount of twist produced in them will be in the proportion 1, $\frac{1}{16}$, $\frac{1}{81}$, &c.; that is, *inversely as the fourth power of the diameter of the wire.* Lastly, taking wires all of the same diameters and lengths, but of different materials, and applying to them the same twisting moment, *the amount of twist will be inversely proportional to the number which we call the modulus of rigidity of the material.* (See Art. 87, and Table III.) The exact rule found experimentally is this:—To find the angle of twist in a brass wire 20 inches long, 0·1 inch diameter, when a twisting moment of 4 pound-inches is applied, multiply 4 by 20, and by a constant number 583·6; divide by the modulus of rigidity, for brass 3,440,000, and by the fourth power of ·1, which is ·0001, and we get the angle of twist 135·7 degrees.

It will be seen from this that the strain is a shear strain. Consider M H G (Fig. 44) to be a cross section of the wire; then a point which is at H before the twist occurs is found to be at G when there is a twist in the wire, and a point such as P' moves to P, but a point O in the centre of the wire does not move. Now there is no such motion at the fixed place A, Fig. 43, and in each section there is more of this motion the farther it is away from A; in fact, the motion is just as it was in the indiarubber of Fig. 37, only that it varies in the section, the motion being greatest at the outside of the wire, and nothing at the centre. The material breaks when the shear stress at the surface becomes too great, and the rule found by experiment is that for any material, what-

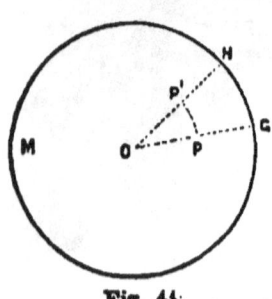

Fig. 44.

Chap. IX.]   TWISTING OF A ROUND SHAFT.   95

ever the length of the wire, the twisting moment which will cause rupture is proportional to the cube of the diameter. It is well known that when a shaft is transmitting power, the horse-power transmitted is proportional to the twisting moment or torque in the shaft multiplied by the number of revolutions made by it per minute. **The rule used by engineers** is this:—*Divide the horse-power by the number of revolutions per minute, and extract the cube root; multiply this result by 3·3, and we have the safe diameter for a wrought-iron shaft.* We use the number 2·88 if it is a steel shaft.

92. Consider a little prism, P B (Fig. 45), whose ends lie in two cross sections of a shaft near together, o being the centre of one of the sections, and o' the centre of the other. The twisting strain causes B to move to B',.regarding P as fixed. (The motion is, of course, usually very much less than I have here shown it.) There must then be shearing forces acting on the ends in opposite directions. If T is the **angle of twist** of the shaft per inch of its length, then B O' B' is T multiplied by O O', and if O P or O' B is r, then B B' is r.T.OO', where T is an angle measured in radians (see Angle, in GLOSSARY). The shear strain in the little prism is B B' divided by P B or O O', so that it is r T, hence the shear stress is N r T (see Art. 87). If a is the area of the end of the little prism in square inches, the shear force acting on it is N r T a, and as this acts in the direction B B' at right angles to the radius, its moment about O O' is N r^2 T a. But we have a similar moment for every such little area into which the cross section may be divided, and to find the total torque we must take the sum of all such terms. Now N and T are the same everywhere, so that in taking such a sum our only difficulty is with the

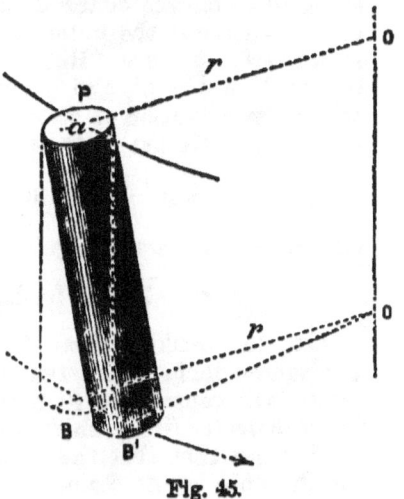

Fig. 45.

factors $r^2 a$. But the sum of all such terms as $r^2 a$ is called the *moment of inertia* of the section about the axis o o', and it has been calculated for us. Thus, if D is the diameter of a round shaft, the moment of inertia of its section about an axis through its centre *at right angles to the section* is $\pi D^4 \div 32$, and for a hollow shaft whose outside diameter is D and inside diameter d, the moment of inertia is $\pi (D^4 - d^4) \div 32$, and hence we see that the moment M necessary to produce a twist *of* T *radians per inch in a round shaft of diameter* D *is*

$$M = \pi N T D^4 \div 32 \qquad (1),$$

and for a hollow shaft it is

$$M = \pi N T (D^4 - d^4) \div 32.$$

Of course, one radian per inch is the same as 57°·2958 per inch.

93. The strength of a shaft is to be calculated on the assumption that rupture occurs when the shear stress N r T mentioned above exceeds the greatest shearing stress to which the material ought to be subjected, and this occurs when r is the outer radius of the shaft or ½ D; that is, when $f = \frac{1}{2}$ N D T. But from equation (1) we find that N T is $32 M \div \pi D^4$, and hence $f = \frac{1}{2} D \times 32 M \div \pi D^4$, and this is the condition of strength of a cylindric shaft. It is more compactly put in the form—

$$M = \frac{\pi D^3 f}{16} \text{ for solid cylindric shafts,}$$

and in the same way we get

$$M = \frac{\pi (D^4 - d^4) f}{16 D} \text{ for hollow cylindric shafts,}$$

f being the breaking shear stress of the material in pounds per square inch, M the twisting moment in pound-inches which will cause rupture, D the outer diameter, and d the inner diameter (if the shaft is hollow) in inches.

We see then that the strength of a solid shaft depends on the cube of its diameter, whereas its stiffness depends on the fourth power of its diameter.

94. The above demonstration is found to agree with experiment, but its results must not be applied except to shafts which are **circular in section.** Our assumption, which experience warranted, was that when such a shaft as A B, Fig. 46, is fixed at B, and when to an arm, C D, a twisting couple is applied, **every straight line in a section remains straight and moves through the same angle as every other line.** But it can be shown that this is not the case for a shaft of any other than a circular section. Thus, let o (Fig. 47), be the centre of gravity of the section P Q S, and let us suppose that a shaft of this section is subjected to the sort of strain I have described. The

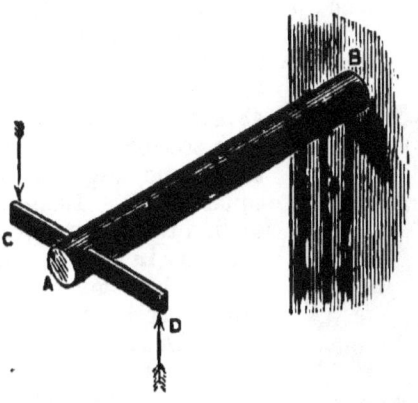

Fig. 46.

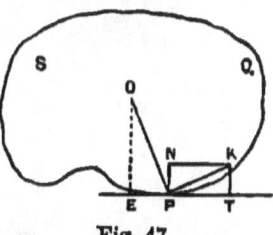

Fig. 47.

shear strain at the point P is in the direction P K perpendicular to O P. Let its amount, which we know to be OP × angle of twist, be represented by the length of P K. It is easy to show that this is just the same as a shear strain, P N, in the direction P N, normal to the surface of the shaft at P, together with a shear strain in the direction P T, tangential to the shaft at P. But shear strain in any

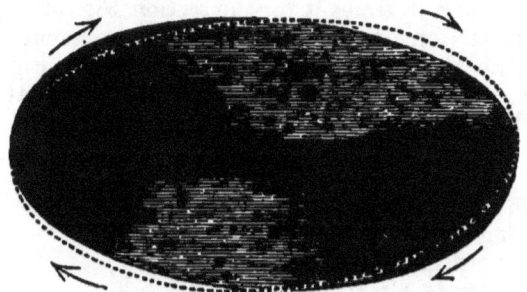

Fig. 48.

direction is always accompanied by a similar strain in a plane at right angles to this direction (see Art. 87), so that since we have the shear P N, we must also have a shear parallel to the axis of the prism along the surface at P, and this cannot be produced merely by a twisting moment. We must imagine that along with the twisting moment there is a force distributed over the surface of the shaft to produce the above effects. The result of an exact investigation is that a twisting couple produces a greater twist than might appear from what I have said in Art. 92, and it also produces a warping of the naturally plane sections of the shaft. Thus Fig. 48 is the shape assumed by each section of an elliptic shaft, and Fig. 49 of a square shaft. Imagine a section to be distinguishable, say in a glass shaft, by a thin layer of a different colour from the rest. Deeper shading indicates greater distance from the observer who is looking towards the fixed end of the shaft. The arrows show the direction of the twisting moment. In the follow-

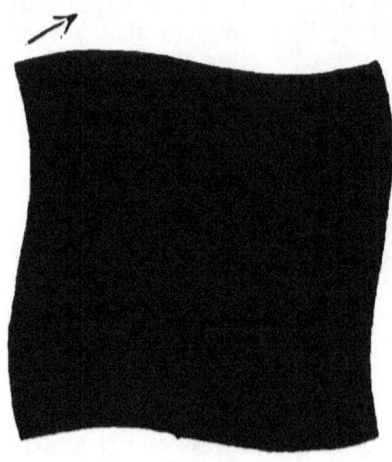

Fig. 49.

ing three sections, instead of the torque for a twist of one radian being equal to N times the moment of inertia of the cross section, it is only ·84 times this for a square section (Fig. 50), ·54 times it for the section Fig. 51, and ·6 times it for the section Fig. 52. Indeed, the square section has only ·88 times the torsional rigidity of a cylindric shaft of the same sectional area; Fig. 51 has ·67 times, and Fig. 52 has ·73 times the torsional rigidity of a cylindric shaft of the same sectional area.

95. A very interesting result of the investigation is that there is always **greatest distortion at that part of the surface** of a shaft

Fig. 50.

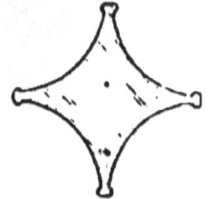

Fig. 51.

(if it has not a circular section) where the surface is **nearest the axis.** Thus, in an elliptic shaft the substance is most strained at the ends of the shorter diameter of the section. Imagine a very light box to be made so as to contain frictionless liquid exactly of the shape of a shaft. If we give a sudden turn to the box about the axis, the liquid will be left behind if the box

Chap. IX.] ELASTIC STRENGTH AND PREVIOUS HISTORY. 99

is circular in section, but it will have motions relatively to the box which can very readily be imagined if the shaft is not circular in section. Now the actual velocity of the liquid at any place relatively to the box is in the same direction as, and is proportional to, the shear in a similar shaft when it is twisted; this has been proved by Sir William Thomson. You will see from this that there is very little strain at the projecting ribs of the shaft, whose section is shown in Fig. 51 and just at the projecting angles of Figs. 50 and 52. This reminds me of a general remark which I have to make, and which I must leave without proof. A solid of any elastic substance cannot experience any finite stress or strain in the neighbourhood of a projecting point, unless acted on by outside forces just at the point. In the neighbourhood of an edge it may have strain only in the direction of the edge, and generally there will be exceedingly great strain and stresses at any re-entrant edges or angles. An important application of the last part of the statement is the well-known practical rule, that every re-entering edge or angle ought to be rounded to prevent risk of rupture in solid pieces designed to bear stress. An illustration of the principle is the stress at the centre of the circular outline in the three sections of shafts, Figs. 53, 54, and 55. In Fig. 53 at o there is no stress when the shaft is twisted; in Fig. 54 the stress may be calculated; in Fig. 55 the stress is exceedingly great for even the smallest twist.

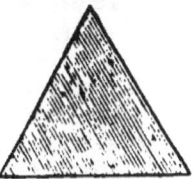

Fig. 52.

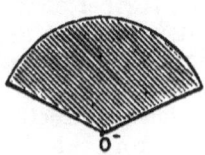

Fig. 53.

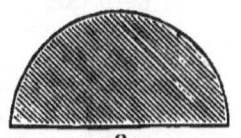

Fig. 54

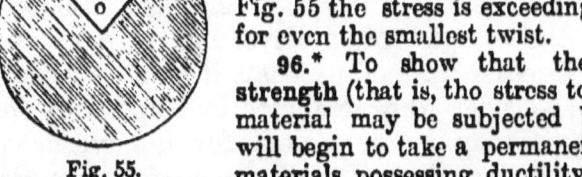

Fig. 55.

96.* To show that the elastic strength (that is, the stress to which a material may be subjected before it will begin to take a permanent set) of materials possessing ductility, like ordinary metals, may vary to a great extent according to the state of strain which exists in the materials when

---

* This paragraph is an abstract of a paper published by Professor James Thomson in the Cambridge and Dublin *Mathematical Journal*, November, 1848.

no external loads are applied to them, let us take a round bar of wrought iron perfectly annealed so that it has no internal strains when unloaded. Now let it bo twisted until the outer portions get permanent set, and continue the twisting process until each portion of the material, from circumference to centre, is strained beyond the limits of elasticity. When this is the case, the stress at every place is the same as at every other, and it is easy to prove that the total resultant couple or torque of resistance at a section is one-third greater than it was when only the outer portions had acquired this stress. If now the twisting load be removed, it will be found that the outer parts of the bar are subjected to negative strain, and the inside to positive strain. If now the bar be subjected to any twisting load less than that which we removed, and in the same direction, it will undergo no further change, will exhibit no further permanent set, and yet this load may be greater than the load which originally produced a permanent set. It will, however, be found that half this load, if it twists the bar in the opposite direction, will produce permanent set. In fact, **the bar in its new state has twice as much elastic strength to resist torsion in the one direction as in the other.** It has two limits of elasticity for opposite kinds of twisting loads, and, if we are to avoid a new permanent set, we must take care that our twisting loads do not exceed these limits. Similar principles operate in regard to beams, and in general the effects are more observable than in shafts subjected to torsion. We can in this way explain why it is that a new beam or shaft takes **a permanent set with even small loads**, since the process of manufacture may have given to portions of the material certain strains which they retain when the beam or shaft is unloaded externally.

97. A shaft is usually subjected to both **bending and torsion** at the same time. The bending is due to its own weight, the weight of pulleys and wheels, and their driving forces. In a crank shaft it is of especial importance to consider the combined effect, but in ordinary shafting it may be neglected, for the reason that in designing a long line of ordinary shafting we really pay more attention to stiffness than to strength. Thus it will be found that the practical rule given in Art. 91 allows a considerable margin of safety as far as mere strength to resist torsion is concerned. In a long line of shafting, if the power is given off at various places with some irregularity, it may even become evident to the eye that the shaft is perpetually twisting and untwisting, for of course the twist is proportional to the horse-power transmitted if the speed is constant. When this is the case, although the shaft may be strong enough, it is not stiff enough. A very long shaft sometimes gets into a state of torsional vibration just in the same way that the cage-rope of a coal-mine gets into a state of longitudinal vibration, due to this want of stiffness. The nature of this vibration will depend on accidental causes, and should the impulses that give rise to it happen to repeat themselves at proper intervals, the vibration may go on increasing until the torsion at some place may be sufficient to produce rupture. In the same way a number of men walking from side to side of a large ship, just taking as much time in going from one side to the other as the ship takes to make a vibration, may make the rolling dangerous.

CHAPTER X.

BENDING.

98. In Fig. 56 C D is a **beam** carrying a weight. We know that the beam transmits the weight to the walls, and that in doing so the beam is kept in a strained condition; we must consider what is the state of strain in the beam. To observe this it will be well to take a beam which is very visibly strained, a beam of india-rubber.

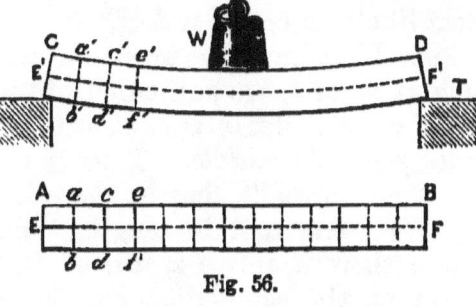

Fig. 56.

A B is its appearance when lying on the table, and we draw upon it a number of parallel lines in chalk or pencil, $a\ b$, $c\ d$, $e\ f$, &c. Now if you support the beam at its two ends, and load it, you will find that the lines $a\ b$, $c\ d$, &c., remain straight, but they are no longer parallel. You will find the distance $a'\ c'$ to be less than $a\ c$, but $b'\ d'$ is greater than $b\ d$. In fact, $a'\ c'$ is compressed, $b'\ d'$ is extended. You will find also that along the line E F there is neither compression nor extension. E' F' remains of its old length, although it is no longer straight. If you consider each cross section of such a beam you will see that the upper part of it is in compression, the lower part of it is in extension, and there is a place in the middle where there is neither compression nor extension. Fig. 57 shows a magnified drawing of the small portion of the beam between two such cross sections. $a\ c\ d\ b$ shows its original shape, $a\ c'\ d'\ b$ is its shape when strained. Evidently there is more compression at $a\ c'$ than at $l\ m'$. The compression becomes less and less as we come nearer G H, then the extension begins, and becomes greater and greater as we get farther away from G H until we get to $b\ d'$, where it is greatest. If the material is likely to break in compression it will be most likely to break at $a\ c'$. If it is likely to break in tension it will be most likely to break at $b\ d'$.

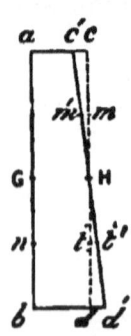

Fig. 57.

99. If we know the tension or compression at any place, such as m' or l', **we can calculate** what it is at any other place, for *the strain is evidently proportional to the distance from the middle*. Thus if at c' there is a compressive strain of ·002, that is, there is a compression of ·002 foot for every foot in length, then half-way between H and c' there is only a strain of ·001. There is the same strain at the same distance below H, but it is now an extension. The material resists being strained in this way, and the pushing and pulling forces which it exerts at the section $c'\ d'$, Fig. 56, are just the forces required to balance all the other forces acting on the part c' D F' T d'.

100. As c' D T d' is a body kept at rest by forces, and which is no longer altering in shape, it is to be regarded as a rigid body.* Now what is the condition under which it is kept at rest? The forces acting on it must satisfy two conditions: 1st, *they ought to balance, if they all act at one point;* 2nd, *their turning moments about any point must balance.* It is this second condition which is more important just now; the turning moments of all the pushing and pulling forces on section c' d' about any point, and it is convenient to take H as this point, must balance the turning moments of all other forces applied to c' D T d'. Now we know what these other forces are; they are the weight W, the weight of c' D T d' itself, and the supporting force at T; their resultant effect is called the bending moment at H. You will generally find that the forces acting on the section are not all mere pushing and pulling forces. Thus in the model, Fig. 58, which shows a beam fixed at one end and loaded at the other; part of the material has been removed, and instead of it we have inserted a chain or link A, which is only capable of exerting

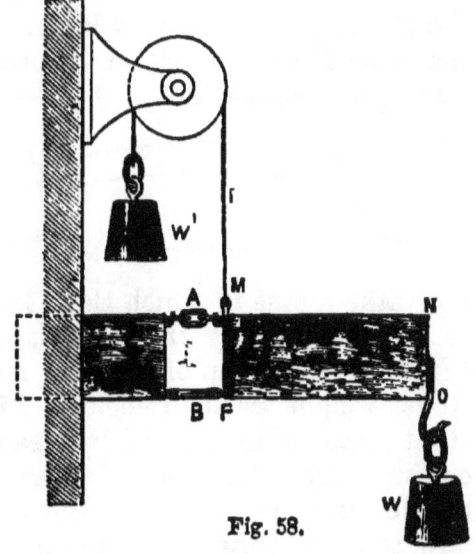

Fig. 58.

---

* In books on mechanics you may have read much about rigid bodies and the laws of their equilibrium, and you may have thought that such bodies had no existence; but you must remember that we can regard a quantity of water, or a piece of steel spring, or a rope, as a rigid body for the time being, if it is being acted on by forces, and is *no longer changing its shape.*

a pull, and a rod B, which is only capable of exerting a push. It is found that these two forces acting on M N O F are not sufficient to keep it at rest; we also need an upward force at M, which is equal to the weight W, together with the weight of M N O F itself. We see then that at such a section as M F of a beam we need pulling and pushing forces, but also to satisfy the first condition given above we need what is called a **shearing force** at M F. At M F the bending moment is W × O F, together with the weight of M N O F × the distance of its centre of gravity from M F. This is to be balanced by the pull in the chain A or the push in the rod B, for these are equal, multiplied by the distance between their lines of action. If a beam is long, the shearing force exerted by the material at a section of the beam is usually not so important as the pushing and pulling forces, and in many cases it is neglected. When a beam is very short the shearing force becomes more important.

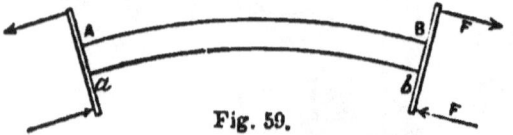

Fig. 59.

101. We will now take a case in which there is only bending moment to be balanced by the material at a section. Let A B (Fig. 59) be a strip of wood or metal originally straight, whose weight we shall neglect. Fix or solder to the ends stout pieces of metal, and by means of cords and weights, or in any other way, exert couples on these ends. Consider now the equilibrium of any portion, say C D B (Fig. 60). At the section C D we know that

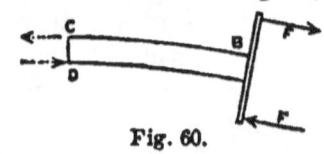

Fig. 60.

pulling and pushing forces must be exerted by the material which exists at the left of C D on the material which exists at the right of C D, and the moments of these just balance the moment of the forces F and F, and this is evidently the same at any section of the strip. The bending moment at any section is then the moment of the couple acting at either end. Let us suppose this to be 20 pound-feet. Magnifying the section c d, as in Fig. 61, and representing the amounts of the pulling and pushing stresses by arrows, we see that as the sum of all the forces one way must be equal to the sum of all the forces acting the other way, and as the stress at each place is proportional to distance from o, the part where there is no stress is a line through o at right angles to the paper (called the *neutral axis*), and this must pass through the centre of gravity of the section.

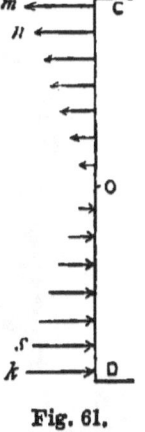

Fig. 61.

102. One has an instinctive feeling that this must be true, but it is difficult to prove it without algebra. If p lbs. is the tensile stress per square inch at the distance one inch above o (or rather the line through o at right angles to the paper), then at the distance x inches above o the tensile stress is px lbs., and at the same distance beneath o there is a compressive stress of p x lbs. If each little strip of area is multiplied by the pressure upon it per square inch, the sum of all the tensile forces ought to be equal to the sum of all the compressive forces. Thus, if a square inches is a very small area at the distance x inches from o, then the sum of all such terms as $p x a$ for places above o ought to be equal to the same sum for places beneath o. Hence, as p is a constant multiplier, the sum of all such terms as $x a$ for places above o ought to be equal to what it is for places beneath o. But this is neither more nor less than saying that the centre of gravity is neither above nor beneath o. In fact, the line through o at right angles to the paper, that is, **the neutral axis of the section, passes through the centre of gravity of the section.**

103. Now the force on any little portion of area is proportional to its distance above or beneath o, and hence the turning moment of this force about o is proportional to the square of this distance, but if *every little area of a section is multiplied by the square of its distance from the neutral axis and the results added together*, we get what is called the **moment of inertia** of the section, hence *the bending moment at a section* *is equal to the stress at one inch from o multiplied by the moment of inertia of the section.*

The moment of the force $p\,x\,a$ which acts on the area a about the neutral axis of the section is $p\,x\,a\,\times\,x$, or $p\,x^2\,a$, and the sum of all such terms for the section is p multiplied by the moment of inertia of the section. Hence, suppose we know the bending moment M at a section, and the moment of inertia I of the section, about the neutral line through its centre of gravity, then $M \div I$ is the stress at one inch from the neutral line. If the extreme edge of the section is y inches from the neutral line, then $y\,\frac{M}{I}$ is the greatest stress there, and must not exceed the breaking stress of the material. $I \div y$ has been calculated for a great number of sections of beams, and it is called the **strength modulus** of the section. The bending moment at a section divided by the strength modulus must not exceed the breaking stress of the material. The strength modulus for a rectangular section is $\frac{1}{6}\,b\,d^2$ where b is breadth and d depth of the section. For a circular section the strength modulus is $\cdot 0982\,d^3$ if d is its diameter. For a hollow circular section it is $\cdot 0982\,\frac{D^4-d^4}{D}$ if D is outside and d inside diameter. The strength modulus is exactly the same for a hollow rectangle (Fig. 62) as for the section (Fig. 63), being $\frac{B\,D^3-b\,d^3}{6\,D}$, where B and b are the outside and inside breadths, and D d the outside and inside depths, of the rectangles forming the section.* For a given amount of material we have the best arrangement for strength in either the I form, or hollow rectangle, since here the material is found where most required, namely, in the top and bottom members, thus giving a greater moment of inertia than if collected near the neutral line.

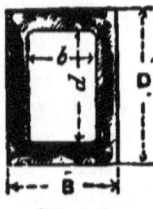
Fig. 62.

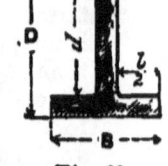
Fig. 63.

104. If, then, we would know what is the amount of the stress everywhere in a section, it is necessary to find the stress at one inch from o by dividing the bending moment at the section by the moment of

---

* A very complete list of these strength moduli will be found in Wertheim.

inertia of the section. Fortunately this moment of inertia has been calculated for us in several cases. To find it, *if the section is rectangular, multiply one-twelfth of the breadth by the cube of the depth; if the section is circular, multiply the fourth power of the diameter by* 0·0491. To take a numerical case, suppose the bending moment to be 8,000 pound-feet at the rectangular section of a beam whose breadth is 2 inches and depth 3 inches, the moment of inertia is $2 \times 3 \times 3 \times 3 \div 12$, or 4·5, hence the stress at one inch from the middle of the depth would be $8,000 \div 4·5$, or 1,778 lbs. per square inch; and the stress at the top or bottom surfaces, which are 1·5 inch from the middle of the section, is $1,778 \times 1·5$, or 2,667 lbs. per square inch; in one case being a tensile, and in the other a compressive stress. If the section of the beam is such that the centre of gravity and neutral line are nearer the bottom than the top, then at the bottom there never can be as great a stress per square inch as at the top. Sometimes a material is such that it can bear much more compressive than tensile stress—cast iron, for instance. In the case of cast iron the section is made so that the centre of gravity is nearer the edge which is to be subjected to tension, in order that the tensile stress may never be so great as the greatest compressive stress. (See B, Fig. 69.)

105. In Fig. 59 the neutral line which passes through the middle of every section, being neither extended nor compressed, is of the original length of the strip. Suppose it to be 1 foot long. When the beam is bent as in the figure, A B becomes longer than this, and $a\ b$ shorter, yet their ends are in the same planes A a and B b. Thus the strip may be considered as a bundle of fibres lying in arcs of circles which have the same centre and subtend the same angle at that centre. If we know their relative lengths we can tell where the centre of the circle is. Now we know the stress per square inch on a certain fibre, and we know its original length, hence we can calculate its present length (see Arts. 49 and 51),

and its length is to the length of the neutral fibre as its radius is to that of the neutral fibre. In this way we find that *the radius of the neutral fibre is numerically equal to the modulus of elasticity of the material multiplied by the moment of inertia of the section, and divided by the bending moment at the section.*

$\frac{M}{I}$ is the stress at one inch from the axis (see Art. 103), and a fibre all along the rod at one inch from the axis is extended proportionally to this stress. Its old length was one foot, therefore its extension of length is $\frac{M}{I} \div E$, a fraction of a foot (see Art. 49), if E is the modulus of elasticity of the material. Every fibre forms the arc of a circle. Now let r inches be the radius of the circle formed by the middle fibre, which is not strained, then $r + 1$ inch is the radius of the fibre we have been considering, and as their ends are in the same radii, we know that the lengths of the fibres are proportional to their radii. The length of the unstrained fibre is 1 foot, and that of the extended one is $1 + \frac{M}{EI}$ feet, hence

$$r : r + 1 :: 1 : 1 + \frac{M}{EI},$$

from which we find that $r = \frac{EI}{M}$, the rule given above. But it is sometimes more convenient to put it in the form $M = \frac{EI}{r}$, or $M = EI \times$ curvature of the strip, rod, or beam. (See GLOSSARY for definition of Curvature.) Now if the strip in its natural unstrained condition had been curved, instead of being straight, you would have found in exactly the same way that $M = EI \times$ change of curvature, or $M = EI \left(\frac{1}{r} - \frac{1}{r_0}\right)$, if r_0 was the radius of curvature of the strip at any place when unstrained, and r is its present radius of curvature.

Example.—A straight strip of tempered steel, 0·7 inch broad, 0·1 inch thick (this represents the *depth* of a beam), is subjected to a bending moment of 100 pound-inches: **find its radius of curvature.** Answer: the moment of inertia of the section is $0·7 \times ·1 \times ·1 \times ·1 \div 12$, or ·0000583. The modulus of elasticity of steel is, say 36,000,000, and $36,000,000 \times ·0000583 \div 100$ gives 21 inches for the radius of curvature of the bent strip.

106. Elastic Curve.—If you take a straight uniform strip of steel and subject it to two equal and opposite

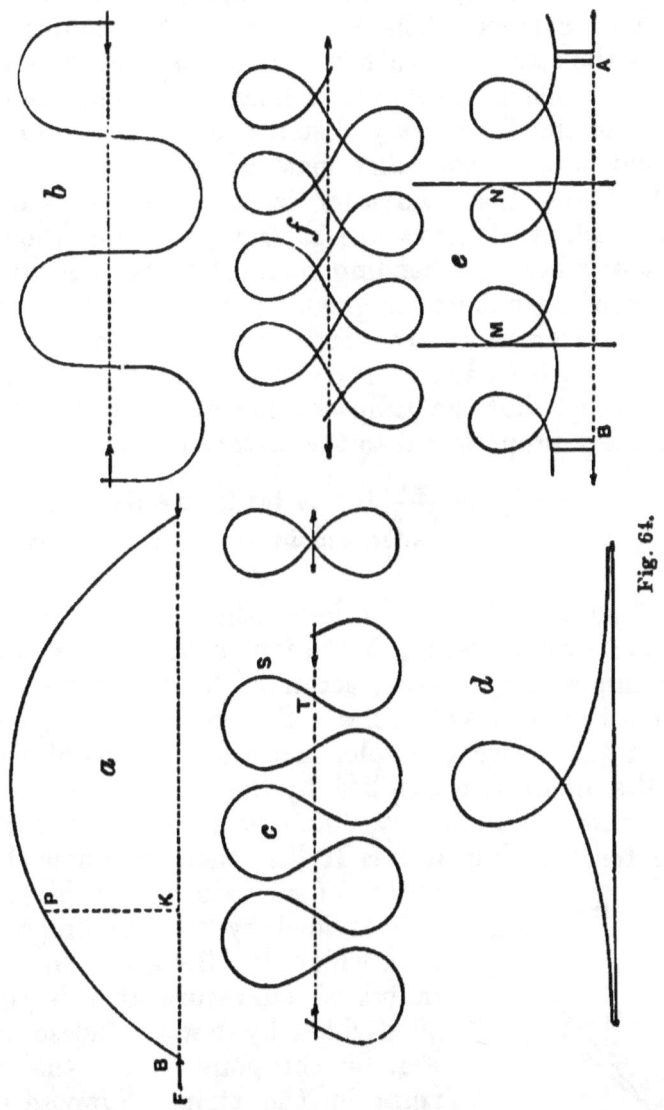

Fig. 64.

forces in the same straight line, the strip will assume one of the forms shown in Fig. 64, which all go under a common name—the *elastic curve*. Now consider the part

P B, Fig. 64 a. Neglecting its own weight, it is acted on by a force F a tB, and at P there must be a force or forces produce balance. There is a force at P tending to compress the steel, but what is of more importance is the fact that F produces a bending moment at P, and the amount of it is the force × the distance P K. Now our strip is everywhere of the same material and section, and the only thing that can alter is its radius of curvature. This radius of curvature at any place we know to be greater when the bending moment is less, and less when the bending moment is greater; in fact, the radius of curvature is inversely proportional to the bending moment, and this really comes to the fact that the radius of curvature at any place P is inversely proportional to the distance P K.

$r = \frac{EI}{M}$, or $\frac{EI}{F \cdot PK}$, if F is the force acting at B. Now E I and F do not alter, and hence $r \times PK =$ some constant number.

You can obtain the shapes shown in Fig. 64 in two ways: first, by taking a straight strip of steel and performing the operation; secondly, by drawing the curves in a series of arcs of circles. Suppose we have calculated, as in the above example, that the modulus of elasticity of the material multiplied by the moment of inertia of its cross-section is, say, 200, and suppose we know that the force acting at B is 10 lbs., then we know that the radius of curvature at P is equal to 200 divided by the bending moment at P, which is 10 × P K. In fact, the radius of curvature at P is equal to 20 divided by P K. Choose now in Fig. 65 the point C as the middle point in the strip. Suppose C D to be 4 inches, then the radius here is 5 inches. Take C O = 5 inches, and with O as centre describe a small arc, C E. Join E O and produce it. Now at E measure E F, and suppose you find it 3·4

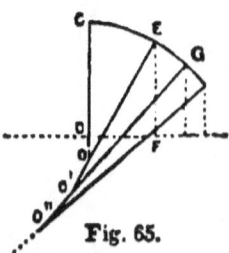

Fig. 65.

inches; divide 20 by 3·4, and we get 5·88 inches, and set this new radius off from E to O'. Take O' as a new centre, and describe the short arc EG of any convenient small length, and in this way proceed until the curve is finished. This is not a very accurate method of drawing the curve unless the arcs are very short, and small errors are apt to have magnified evil effects, but I know of no better exercise to impress upon you the connection between radius of curvature of a strip and the bending moment which produces it. You are therefore supposed to have actually drawn one such curve at least before proceeding with your study of this subject.

107. Parts of these curves happen to be the shape taken by liquids, because of their capillary action, between two solid plates. They are also the shapes of the arches which are best fitted to withstand fluid pressure. Thus, for instance, in Fig. 66 the curved water from M to N is of the shape of the curve Fig. 64 e, from M to N, the free water level being the line AB; and in Fig. 67 the middle line of the joints of the arch M to N is the same curve inverted. The water, whose pressure it resists, has as free water level the line AB.

Fig. 66.

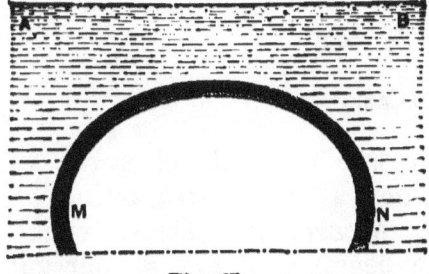

Fig. 67.

108. When a strip of elastic material is bent, it not only alters its shape in the well-known way, but it alters the form of its cross-section. On the convex side of the strip the breadth becomes concave, and on the concave side of the strip the breadth becomes convex. It is very easy to try this for yourself on a broad strip of steel or a bar of india-rubber. These saddle shapes of the surfaces

are due to the fact that when each fibre is pulled it gets thinner as well as longer (see Art. 56), and when it is pushed it gets broader as well as shorter, and it is very curious that this action should not interfere perceptibly with the laws of bending as I have given them to you.

108a. It may now be interesting to consider the **relation between bending and twisting.** We have seen (Art. 105) that a couple, M, applied to produce bending gives a curvature $\frac{M}{EI}$. Now I, the moment of inertia of a circular section about a diameter, is $\frac{\pi D^4}{64}$, hence the curvature is $\frac{M}{E} \div \frac{\pi D^4}{64}$. We have also proved (Art. 93) that a couple, M, applied to produce twisting in a cylindric shaft, gives an angle of twist $\frac{M}{N} \div \frac{\pi D^4}{32}$.

Now, if, in the above cases, the bending and twisting couples are equal, we have

$$\frac{\text{Angle of Twist}}{\text{Curvature}} = \frac{E}{2N} = \frac{1}{2a} \div \frac{1}{2(a+b)} = \frac{a+b}{a} = 1 + \frac{b}{a}.$$

Experiment shows that for isotropic substances the ratio $\frac{b}{a}$ lies between 0 and $\frac{1}{2}$. Hence, $\frac{E}{2N}$ is always greater than unity. Therefore, twist must always exceed bending when the couples producing them are equal.

CHAPTER XI.

BEAMS.

109. To be able to calculate the state of strain of a beam it is necessary to know all the forces acting on it from the outside; these are the loads, which include the weight of the beam, and the supporting forces at its ends. If we know the loads, it is easy to calculate the supporting forces when a beam is supported at the ends. The load may be concentrated at one or more points, or it may be distributed uniformly over the whole or part of the beam.

It may be a *dead* load or a *live* load. A dead load is one which has been applied very gradually, and remains pretty much the same for a long time; a live load is one which has been more or less suddenly applied. Given the loads, we are always able to determine the supporting forces if there are only two, either by the ordinary rules of mechanics or by the graphic method described in Art. 143. When we know the necessary supporting forces at the ends of a beam, we can take care that there are suitable means of support for the beam.

110. Methods of Supporting Beams.—In practice, whenever it is possible, the ends of a beam are not merely supported, but they are fixed by being built into brickwork or masonry, for it is known that fixing the ends strengthens and stiffens a beam very considerably (see Art. 123). Thus also the cross-beams of a railway bridge are well fixed at the ends by means of bolts or rivets. Timber structures are always attached as rigidly as possible to their supports, and this is the case with all structures in which there is no fear of unequal expansion by heat. A long iron beam merely rests upon masonry or timber supports, without being rigidly attached, because the iron expands during the summer and contracts during the winter, more than the timber or masonry, and every facility must be given for relative motion due to these causes. Thus one end of a long iron roof-principal or iron girder generally rests upon a carriage or frame supported on rollers.

111. The supporting force at the fixed end of a beam is often rather indeterminate, but if one end rests upon a carriage we may regard the supporting force there as being nearly perpendicular to the plate on which the rollers rest, and this supposition enables us to find both supporting forces (see Art. 148). In what follows my attention is mostly devoted to beams which are horizontal and are supported by vertical forces at one or both ends.

I

112. When beams carry loads they are not usually subjected to the same bending moment everywhere, and the shearing force is also different at different places. Take any simple case—for instance, a beam A B, Fig. 68, supported at the ends, and loaded uniformly all along its length. If the total load is 2,000 lbs., then each upward supporting force at A and B is 1,000 lbs. Now at any point, C, the bending moment is 1,000 × C B acting against the hands of a watch round C, minus the load on the part C B multiplied by half the distance C B. Erect a perpendicular, C E, and make its length represent, on some scale, the bending moment at C. Do the same for a number of points, and by joining the ends of all the perpendiculars you will get a curve which shows at a glance the bending moment everywhere. In Table IV. the figures M M are diagrams of bending moment which have been calculated by the graphic method described in Art. 143.

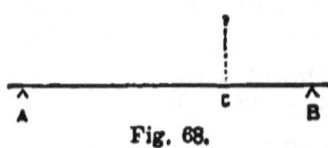

Fig. 68.

When the upper parts of sections of a beam are in compression, the bending moment is usually measured from A B upwards. When the upper parts of sections of a beam are in extension, the bending moment is usually measured from A B downwards. It would have been difficult to give the bending moment in every case to the same scale, as the greatest bending moment in case I. of the Table (p. 116) is twelve times the greatest bending moment in case VI. (p. 118). Hence, if we regard the scale for case I. as 120 pound-feet per inch, the scales for the other cases are 60, 30, 15, 15, and 10 pound-feet to the inch.

113. Again, the shearing force at C (Fig. 68) is simply the upward force at B minus the whole load on the part C B. Set off on the perpendicular at C a distance equal to the shearing force there; do the same for other points, and draw a curve showing the shearing force everywhere. To know the shearing force at every section of a beam is very important in railway girders, because the

lattice-work—that is, the struts and ties which connect the upper and lower horizontal booms—is proportioned to resist the shearing force. It is the same with the thin central web of a wrought-iron girder, if the girder is formed of plates of iron riveted together. But in small beams of cast iron or timber, and even in wrought-iron girders that have been rolled in one piece, the web is usually made so thick that it is unnecessary to try if it is strong enough to resist the shearing force.

114. To resist bending moment, only those parts of a beam which are far from the neutral line are of much importance—hence, in iron beams we have two flanges far apart connected by means of the web or by lattice-work. Thus, in Fig. 69, A is the usual section of a wrought-iron beam, and B of a cast-iron beam. The neutral line o o in each case passes through the centre of gravity of the section. All parts of the section below o o are subjected to tensile stress, all above o o are subjected to compressive stress. Because wrought iron will resist nearly as much tensile stress as compressive before it breaks, the two flanges of A are made equal in area. But inasmuch as cast iron will stand about $4\frac{1}{2}$ times as great compressive stress as tensile, the flange c c, subjected to tension, has about $4\frac{1}{2}$ times the area of d, which is the compressed flange. Thus, the total breaking stress on one of the flanges is equal to the total breaking stress on the other. Suppose the area of a flange is 10 square inches, and its breaking stress is 50,000 lbs. per square inch, then the total breaking stress on this area is 10 × 50,000, or 500,000 lbs. If, now, the distance between the centres of gravity of the two flanges is two feet, we can say that the bending moment which will

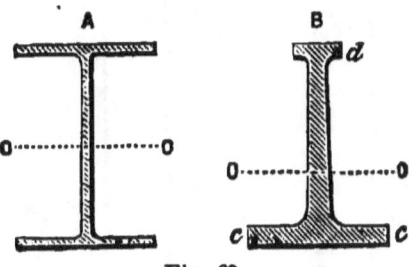

Fig. 69.

TABLE IV.

Nature of Support and Loading.	Sketches of Beams and Diagrams of Bending Moment.	Strength.	Deflection.
Beam fixed at one end. Load w at the free end. Greatest bending moment is at A, and is w multiplied by length of beam A B.	CASE I.	¼	16
Beam fixed at one end. Load w spread all over. Greatest bending moment is at A, and is w multiplied by half length of beam.	CASE II. LOAD W SPREAD ALL OVER. PARABOLIC CURVE.	½	6

TABLE IV. (*continued*).

Nature of Support and Loading.	Sketches of Beams and Diagrams of Bending Moment.	Strength.	Deflection.
Beam supported at both ends. Load w in middle. Greatest bending moment occurs in the middle, and is w multiplied by one-fourth of length of beam A B.	CASE III.	1	1
Beam supported at both ends. Load w spread all over. Greatest bending moment occurs in the middle, and is w multiplied by one-eighth of length of beam A B.	CASE IV. LOAD W SPREAD ALL OVER. PARABOLIC CURVE.	2	·625

118   PRACTICAL MECHANICS.   [Chap. XI.

TABLE IV. (*continued*)

Sketches of Beams and Diagrams of Bending Moment.

Nature of Support and Loading.	Sketches of Beams and Diagrams of Bending Moment.	Strength.	Deflection.
Beam *fixed* at both ends. Load w at middle. Greatest bending moment occurs both at the middle and at the ends, and is w multiplied by one-eighth of length of beam A B. There is no bending moment at the points C and D, and these are called points of inflection.	CASE V.	2	·25
Beam *fixed* at both ends. Load w spread all over. Greatest bending moment occurs at the ends, and is w multiplied by one-twelfth of length of beam A B. There is no bending moment at the points C and D, and these are called points of inflection.	CASE VI. LOAD W SPREAD ALL OVER	3	·125

destroy the beam if it acts at this section is 500,000 × 2, or 1,000,000 pound-feet.

115. When much of the material has been left near the middle part of the section, as it is in ordinary timber beams, it is not so easy to make the calculation, for although much of the timber is in a position where it is but little capable of resisting bending moment, yet it does resist to some extent. Again, in iron beams it is usual to shape them everywhere so that those sections where there is but little bending moment to be resisted are made with smaller flanges, or else flanges which are nearer together. If we have a diagram, such as we see in the various cases of Table IV., showing the bending moment at every part of the beam, we simply vary the section, so that it is just capable of resisting the bending moment which acts there. Now, timber beams, as a general rule, are everywhere of the same rectangular section. There is one place, the place of greatest bending moment, where such a beam is likely to break; we therefore calculate the size of the section to withstand this greatest bending moment.

116. Suppose we take a certain beam which has everywhere the same section, and we load it in various ways. Thus, the load may be hung from one end of the beam, the other end being rigidly fixed, say by being built into a wall. When we say that the end of a beam is fixed, we mean that it is rigidly held in position, whereas when we say that a beam is supported at its ends, we mean that it is merely held up there. In Table IV. six ways are shown in which the same length of beam is supposed to be loaded. The total load is supposed to be the same in every case, and the length from A to B is supposed to be the same. Then, we see that when the beam is fixed at both ends, and the load spread over it, it is 12 times as strong as when one end is fixed, and the whole load hung from the other end. This means that if, with the beam fixed at one end, a load of one ton, hung at the other end, breaks the beam, then, when fixed at both ends, and the load spread uniformly over it, the same-sized beam will carry 12 tons. Hence, if experiments are made on the strength of the beam when loaded in any of these ways, we know what its strength ought to be when loaded in any of the other ways. Now, a great many experiments have been made upon beams of rectangular section, supported at both ends and loaded in the middle, the third case given in the Table; and from these experiments we know how to find the load which such a beam will carry. Having found this, we know that when loaded and supported in a different way, the beam will carry more or less according to the numbers in the column headed *Strength*. The rule which has been deduced from experiments on beams whose sections are rectangular is this:—

A beam supported and loaded in any of the ways shown in Table IV. will break with a total load which is found by *multiplying together the breadth of the section in inches, the square of the depth in inches, the number called strength in Table IV., the number called strength in Table VI., and dividing the product by the length of the beam,* A B, *in feet.*

Example.—A beam of English oak, 20 feet long, 9 inches broad, 12 inches deep, is *fixed* at the ends. What load placed in the middle will break it? This is case V. of Table IV., and the relative strength is given as 2 in the same Table. Opposite English oak, in Table VI., we find the number 557; and hence, $9 \times 12 \times 12 \times 2 \times 557 \div 20$, or 72,188 lbs., or more than 32 tons, is the answer.

117. Suppose the breaking load on a beam of timber is found to be 32 tons, you would follow the usual practice if you really never placed on it a load of more than 8 tons. Thus, you divide the breaking load by 4 to get the safe load, or the working load. This number 4 is called a **factor of safety**. The usual factors of safety employed in structures generally are given in the following Table:—

TABLE V.
Factors of Safety.

Material.	A Dead Load, or one that does not alter.	A Live Load, or one that alters.		
		In Temporary Structures.	In Permanent Structures.	In Structures subjected to Shocks.
Wrought Iron and Steel	3	4	4 to 5	10
Cast Iron . . .	3	4	5	10
Timber	—	4	10	—
Brickwork . . .	—	—	6	—
Masonry . . .	20	—	20 to 30	—

118. You must specially remember that it has been found by experience that if we have beams of the same material of rectangular section loaded in the same way, *the strength is doubled if we double the breadth of the beam or halve its length; but if we double the depth, we increase the strength four times.*

TABLE VI.
Beams Supported at the Ends and Loaded in the Middle.

Nature of Material.	Strength.	Deflection.
Teak	820	·00018
Oak	450 to 600	·00044 to ·00020
English Oak . .	557	·0003
Ash	675	·00026
Beech	518	·00031
Pitch Pine . .	544	·00035
Red Pine . . .	450	·00023
Fir	370	·0005 to ·0002
Larch	284	·00041
Deal	600	·00023
Elm	337	·00061
Cast Iron . .	2540	·000024
Wrought Iron .	3470	·000016
Hammered Steel .	6400	·000013

The numbers given in this Table are merely the average values found by various experimenters. You may wish, however, to find for yourself whether they are correct or not. You are designing a beam of pitch-pine, say; then take a rod of pitch-pine, 1 foot long, 1 inch broad, 1 inch deep; support it at the ends, and load in the middle till it breaks; the Table says that the load will be 544 lbs., but you may find it to be more or less than this. **Remember also that it is near the middle that your beam is likely to break;** this, then, ought to be the soundest and most evenly grained part of the timber if possible, and the specimens which you try ought to be as nearly as possible the same kind of timber.

119. When a beam is loaded in any way, you know how to find the bending moment at any place, and if you know the modulus of elasticity of the material, and the moment of inertia of the section, you can find the **curvature of the beam.** You may draw a bent beam, then, in the same way as you drew the springs of Fig. 64, but the beam is so little curved usually that you will have difficulty in getting compasses long enough. In this case it is usual to diminish all the radii in some large proportion, remembering that the deflection of your beam as you draw it is increased in this proportion. For a beam fixed at one end and loaded at the other you would get a curve just like the portion s т in Fig. 64 c, s being the fixed end and т the loaded end.

120. The important thing to know is the **deflection of a beam**—that is, *the greatest yielding of any part of it.*

It can be proved mathematically, from what has been given in Art. 105, that if D is the deflection of a beam whose cross section is the same everywhere, w the load, L the length, I the moment of inertia of the section, and E the modulus of elasticity, and if all these are in inches and pounds, or in any other units so long as they are all in the same units, then

$$D = \frac{W L^3}{3 E I}$$ for a beam fixed at one end and loaded at the other.

$$D = \frac{3}{8}\frac{WL^3}{3EI} \quad \text{for a beam fixed at one end, and loaded uniformly.}$$

$$D = \frac{1}{16}\frac{WL^3}{3EI} \quad \text{for a beam supported at the ends and loaded in the middle.}$$

$$D = \frac{5}{8}\frac{1}{16}\frac{WL^3}{3EI} \quad \text{for a beam supported at the ends and loaded uniformly.}$$

The third of these formulæ is the one most needed. It is by means of this formula that the modulus of elasticity is generally determined. Thus in careful experiments with an iron beam, 1 inch broad, 1 inch deep, carried on supports 24 inches asunder, suppose we find that a load of 2,000 lbs. produces a deflection of one-quarter of an inch. Now I for the beam is $\frac{1 \times 1 \times 1 \times 1}{12}$, or $\frac{1}{12}$. The third formula given above becomes $\cdot 25 = \frac{1}{16}\frac{2000 \times 24 \times 24 \times 24}{3 E \times \frac{1}{12}}$, and from this we find that E is 27,648,000 lbs. per square inch.

Again, taking the fifth of the cases shown in Table IV., I find that 560 lbs. produced a deflection of 0·22 inch in a beam of wood 24 inches long, 1¾ inch square. Here $I = 1\cdot75 \times 1\cdot75 \times 1\cdot75 \times 1\cdot75 \div 12$, or ·781, and $\cdot 22 = \frac{1}{16}\frac{560 \times 24 \times 24 \times 24}{3 E \times \cdot 781}$, from which we find that E is 938,656 lbs. per square inch.

Again, from Table VI. we see that a beam of teak 12 inches long, 1 inch broad, 1 inch deep, gets a deflection of ·00018 inch for a load of 1 lb. Here the moment of inertia of the cross section is $\frac{1}{12}$ and $\cdot 00018 = \frac{1}{16}\frac{1 \times 12 \times 12 \times 12}{3 E \times \frac{1}{12}}$, from which we find that E for teak is 2,400,000 lbs. per square inch.

121. Take a small beam, A B, Fig. 70, supported at the ends, and load it in the middle. Measure carefully the **deflection** or lowering of the middle point. This is called the deflection of such a beam. Now this distance will usually be small, and so you had better magnify it by letting the string c w pass over the little axle E, which carries a long pointer. This pointer will show on the scale P K a magnification of the deflection. You will find that the more load you place at c, the greater is the deflection; and in fact that the deflection is proportional to the load until your loads become great enough to produce permanent set, when (Art. 52) the deflections increase more rapidly than the load. If now you use a beam of the same material but of double the breadth, then for the same load you will get one-half the old deflection. If you use a beam of double the depth, then

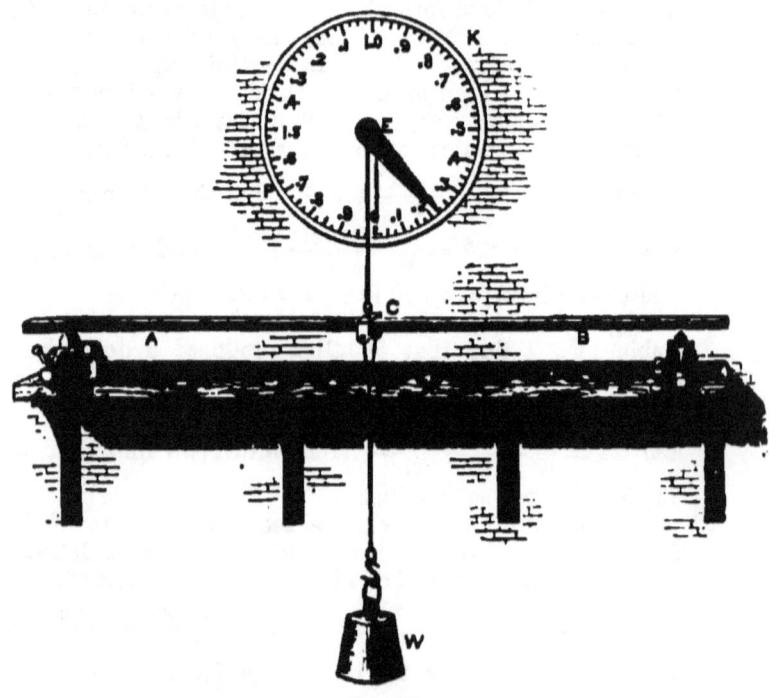

Fig. 70.

for the same load you will get only one-eighth of the old deflection. Also, if you double the length of your beam, using the same load, you will get eight times the old deflection. A very instructive series of experiments may be made very easily in this subject, and you will not thoroughly understand the matter unless you make a few such experiments. It is found that a beam of pitch pine, 1 foot long, 1 inch broad, and 1 inch deep, supported at its two ends and loaded in the middle, is deflected ·00035 inch by a load of 1 lb. This explains the numbers given in Table VI. It is found that if the same beam is fixed at one end and loaded at the other (first case of Table IV.), the deflection is 16 times as great, whereas if the beam is fixed at both ends and the load is spread uniformly (last case of Table IV.), the deflection is only ·125, or one-eighth as great. This explains the "deflection" column of Table IV.

122. The rule, then, to find the deflection in inches of any beam loaded in any of the ways shown in Table IV. is this:—*Multiply together the cube of the length in feet, the total load in pounds, the number called deflection in Table IV., and the number called deflection in Table VI., and divide the product by the breadth of the beam in inches, and by the cube of the depth in inches.*

Example.—A beam 20 feet long, 10 inches broad, 15 inches deep, of pitch pine, fixed at one end and having spread all over it a total load of 4,000 lbs.—what is its deflection? Here the number in Table IV. is 6, and in Table VI. it is ·00035; hence we have 20 × 20 × 20 × 4,000 × 6 × ·00035 divided by 10, and again divided by 15 times 15 times 15, which gives as answer 1·99 inch. The end of the beam would be deflected this distance.

123. A beam is said to be stiff if its deflection is small, and we say that the stiffness of a beam supported and loaded in the various ways shown in Table IV. is for the various cases $\frac{1}{16}$, $\frac{1}{6}$, 1, 1·6, 4, 8. In fact, a beam of a certain length carrying a certain load is 128 times stiffer when it is fixed at the ends and loaded uniformly than when it is fixed at one end and loaded at the other end.

It is well to remember that when we double the breadth of a beam we double its strength and also its stiffness, but if we double its depth we get four times the strength and eight times the stiffness. Beams required to be very stiff ought to be very deep. Care must be taken, however, that they are laterally supported, else they will buckle. If you double the length of a beam you get half the strength, but you only get one-eighth of the stiffness.

124. What about beams that are not rectangular in section? Suppose we have a beam of the same section everywhere, whose strength and stiffness we know, and suppose we want to know the strength and stiffness of another beam which has the same form of section—that is, suppose the new section is such that all the old *lateral* dimensions are increased in a certain ratio—then the strength and stiffness increase in this ratio; if all the old *vertical* dimensions are increased in a certain ratio, then the strength increases as the square of this ratio, and the stiffness increases as the cube of this ratio. The effect of change of length is just the same as it was with rectangular beams, and we know the effect produced by different methods of supporting and loading the beam from Table IV.

> From Arts. 103 and 112 it is evident that the load which a beam will carry without breaking is proportional to the *strength modulus* of its section divided by the length of the beam. The deflection of the beam is proportional to the load multiplied by the cube of the length, divided by the moment of inertia of the cross section.

125. At the Imperial College of Engineering, in Japan, we had a testing machine with which I have made a great many experiments with my students. It increased the load on a beam at a uniform rate, and registered the load and deflection of the beam at every instant—that is, it drew a curve, each point of which showed the deflection and the load which produced it. Mr. George Cawley, instructor in mechanical engineering at the college, lithographed a number of these curves, taken by himself; and although the experiments were made on Japanese wood, so that the actual amounts of load and deflection are not of general interest, yet the shapes of the curves are so interesting as to be

Chap. XI.] STRENGTH AND STIFFNESS OF TIMBER. 127

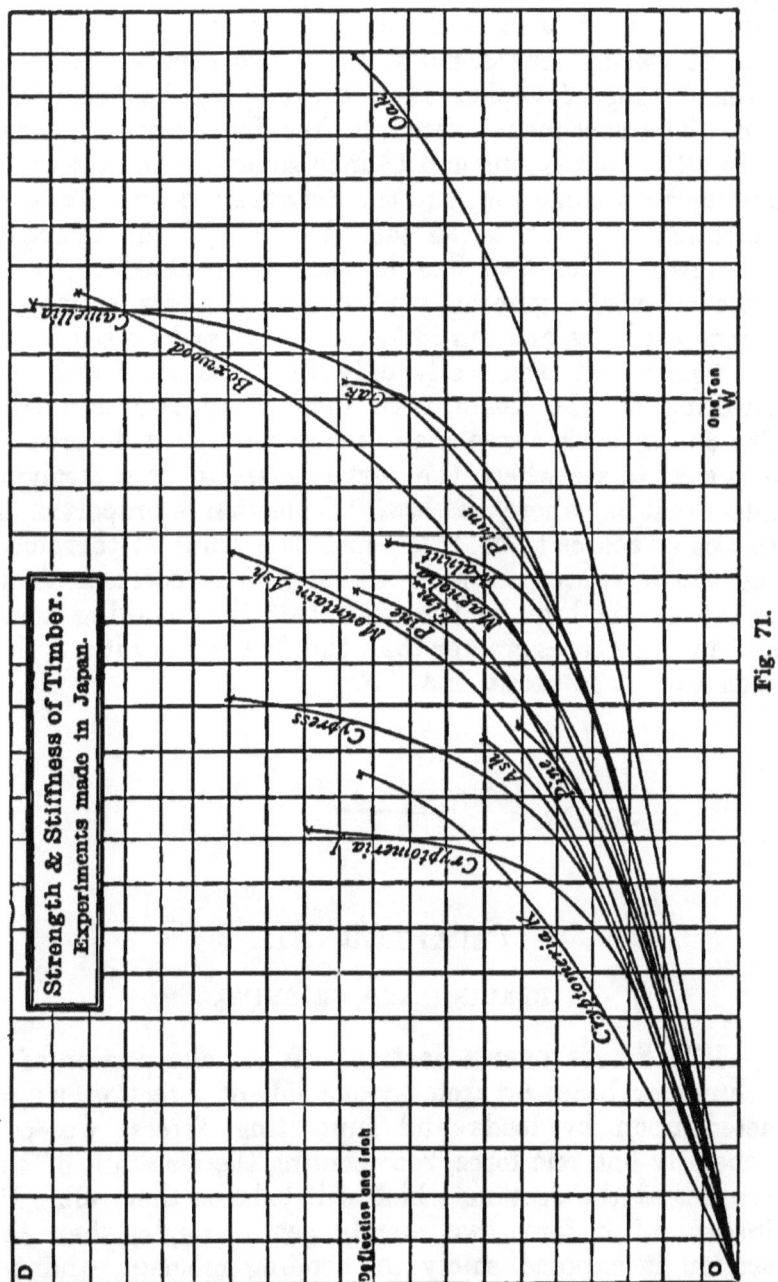

Fig. 71.

worthy of publication. With only one exception, two beams were broken and two curves taken for each kind of wood. The mean of these two curves has been given in Fig. 71—that is, a curve lying between the two. The specimens were all free from knots. They were all 28 inches long and 1¾ inch square. The distance o w represents one ton, and the distance o d represents a deflection of 2 inches, so that the scale of the diagram is known. The load was in each case added to at a uniform rate, beginning with o, and the rate at which it increased was one ton in two minutes, and we see from the figure that practically only in three cases did the breaking of the beam take more than two minutes. The end of each curve shows where the specimen broke; it is easy to see where the curve ceases to be a straight line—that is, where the law, " Deflection is proportional to Load," ceases to be true; and this point is therefore the *elastic limit*. In some cases the load corresponding to the elastic limit is less than half the breaking load, and in some cases greater than this, but usually it may be seen that it is about one-half.

CHAPTER XII.

BENDING AND CRUSHING.

126. Stress over a Section.—When any portion of a column or beam or arch on one side of a section, b c, is acted upon by loads and supporting forces, we can generally find one force, representing the resultant of the stresses at the section, which will balance them all. If, instead of a force, we merely get a couple, then the section is exposed solely to bending moment, and we know now how to find the effect of this. If the force is parallel to the section, then we know that the section is either exposed to mere shearing strain or shearing and bending, as in a horizontal beam with vertical loads; but if the force is inclined to the section, there will usually be shearing and bending, and besides this a uniform distribution of compression or extension all over the section. In practice we generally find that compression and bending alone have to be considered. Thus, if B C (Fig. 72) is the edge view of the section of an arch ring, and if P F is the resultant force in magnitude and direction, and if O represents a line through the centre of gravity of the section at right angles to the paper; then P K the resolved part of P F parallel to B C is the shearing force which must be resisted by the section. F K × O K is the bending moment at the section, causing the parts between O and B to be compressed, and the parts between O and C to be extended. But besides this we must suppose the compressing force F K to be distributed over the whole section. This will increase the compression over the part O B, and will diminish the tension over the part O C, which mere bending would have produced.

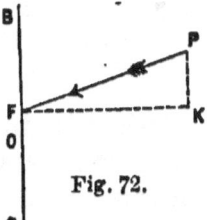

Fig. 72.

127. Thus, in Art. 103 we saw that $O B \cdot \frac{K F \cdot O F}{I}$, where I is the moment of inertia of the section about the axis through O, is the compressive stress at B, due to mere bending moment, and to this we must now add $\frac{K F}{A}$ if A is the area of the section. Hence, the resistance to crushing of the material per square inch must be greater than

$$O B \cdot \frac{K F \cdot O F}{I} + \frac{K F}{A}.$$

Of course the tensile stress at C is

$$O C \cdot \frac{K F \cdot O F}{I} - \frac{K F}{A}.$$

J

If the section is of such a nature that we never wish any portion of it to resist tensile stress, this second expression must be 0, or less than 0. This is usually assumed to be the case in stone or brick bridges, and it is easy to show that if the section is rectangular it leads to the general condition that o F ought never to be more than one-sixth of o B or o c; in fact, that the **resultant force P F must fall within the middle third of every joint of the stone work.** If it falls outside the middle third of the joint, you will have to depend on the resistance of the cement of the joint to tensile stresses, and this is not usually regarded as a safe thing to do.

128. Struts and Pillars.—I disposed much too easily of the compression of a strut in Chap. VI. At short distances from the ends of a bar subjected to pull, the tensile stress is pretty uniformly distributed over the cross section, and whether the bar is long or short the material has nearly as much freedom to get uniform tensile strain in one case as in another. But this is different in struts. If a strut is long it breaks by bending; if it is just so short that we know there is no bending, the load per square inch that will break it may be taken as representing its resistance to crushing; but even this is not such a resistance as a cube of the material would offer. If we take a much shorter column, say a thin disc, the way in which the load is applied may be such as to prevent the lateral spreading which always accompanies compression, and a much greater load is required to crush the material than might have been expected. If a number of specimens of cast iron are taken one-quarter inch square, the first being a cube and the last being 1 inch in length, it will be found that the load which they will support diminishes gradually from 72 tons per square inch to 45 tons. After a certain height is passed the rupture seems to be produced rather by sliding along an oblique section than by mere crushing at a cross section.

129. When a strut or column is of considerable length

it usually bends before it breaks. **Professor Gordon designed a formula** based on this assumption which fairly well represents the results of experiments, and although it is known not to satisfy the facts of the case so well when elastic strength has to be considered, yet it is so easy of application, and is, on the whole, so correct, that I give it in preference to the more correct rule, based on the theory of Euler, which will be found in Professor Unwin's "Machine Design."

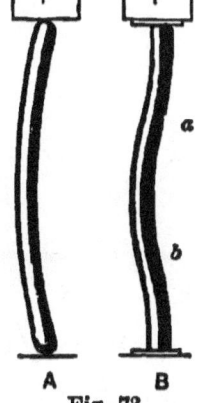

Fig. 73.

Usually the total load divided by the area of cross section is regarded as the stress on the material, but by Art. 126 we see that to this must be added the stresses produced by such bending as the strut undergoes. The result is that the stress on the strut as usually calculated must be increased by a fraction of itself which depends on the square of the length of the strut divided by the moment of inertia of the cross section regarded as the cross section of a beam. The practical rule becomes then—For a *strut whose ends are hinged*, or a *column whose ends are not fixed*, as A, Fig. 73, the breaking load in pounds is equal to the breaking stress per square inch given in Table VII. multiplied by the area of cross section in square inches, and divided by $1 + n$ B where n is given in Table IX. and B is given in Table VIII.

TABLE VII.

	Breaking Stress, in pounds per square inch.
Cast Iron	80,000
Wrought Iron	36,000
Timber	7,200

TABLE VIII.*

Value of B for struts of the sections shown in Table IX. The first column gives the length of the strut divided by its least lateral dimension.

Length divided by Lateral Dimension d.	B for Cast Iron.	B for Wrought Iron.	B for Strong Dry Timber.
10	0·748	0·132	1·
15	1·68	0·300	3·6
20	3·00	0·532	6·4
25	4·64	0·832	10·0
30	6·76	1·200	14·4
35	9·20	1·632	19·6
40	12·00	2·132	25·6
45	18·72	3·332	40·0

TABLE IX.

Values of n for struts and pillars of the following sections:—

Section		Value of n.
Square of side d, or rectangle with smallest side d.		1·00
Hollow rectangle, or square with thin sides		0·50
Circle, diameter d		1·33
Thin ring, external diameter d.		0·66
Angle iron, smallest side d		2·00
Cruciform, smallest breadth d.		2·00

* Modified from Professor Fleeming Jenkin's article on *Bridges* in the "Encyclopædia Britannica."

If we want the breaking load for a strut whose ends are not hinged, it is necessary to find in what way it tends to bend, and to use the above rule regarding the strut as hinged at *two* points of contrary flexure. Thus in Fig. 73 the strut or column B is as strong as a strut hinged or rounded at both ends, whose length is only ab. The rule becomes—For *a strut fixed at both ends*, calculate by the above rule, but take n one fourth of what I have given in Table IX. For a strut, one end of which is fixed and the other is only hinged, calculate the breaking load as if both its ends were hinged, then calculate it as if both its ends were fixed, and take the mean value of the two answers.

130. The Teeth of Wheels.—When toothed wheels drive each other, their teeth tend to break like little beams fixed at one end. It is usual in considering their strength to regard the pressure between two teeth as acting at a corner, because this may accidentally occur, and it is the most trying condition. There are usually two pairs of teeth in contact at once, so we consider that only half the total horse-power has ever to be transmitted by one pair of teeth. This transmitted horse-power, multiplied by 33,000, divided by the circumferential velocity of the wheel per minute, is of course the pressure in pounds which each tooth has to withstand. Imagine the tooth to tend to break at a section making 45° with the depth, just as we know it would break if the corner were struck smartly with a hammer. This consideration leads to the rule, that the pitch is proportional to the square root of the pressure, divided by the greatest safe stress per square inch to which the material may be subjected.

131. Flat Plates.—A plate, round or square, either merely supported or firmly fixed all round its edge, will carry a total load uniformly spread over it which is simply proportional to the square of the thickness, and is not dependent on the area of the plate. Fixing the edges adds a quarter to the strength.

132. Similar Structures Similarly Loaded.—If a girder is loaded mainly by its own weight, then any other girder made to the same drawing but on a different scale would be a similar structure similarly loaded; and this is the name given to all structures made from the same drawings but to different scales, if their loads are in the same proportions to the weights of the structures themselves. It will be found that in all cases the stress at similar places is proportional to the size of the structure—that is, the weakness of the structure is in direct proportion to its size.

This is easily seen if we imagine the structure to be such a simple one as a rod, A, Fig. 74, carrying a weighty

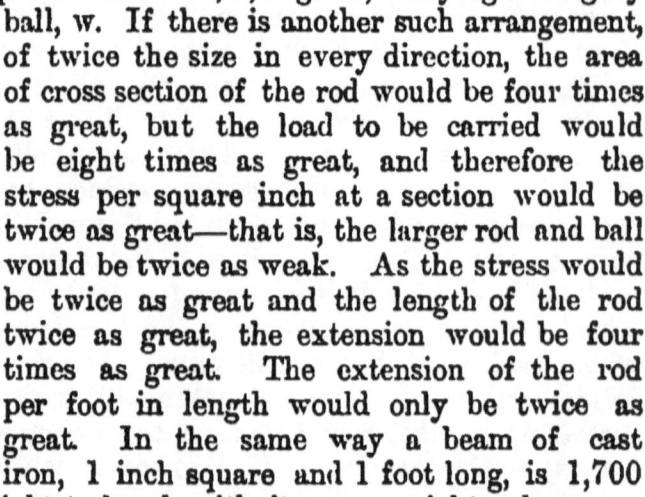

Fig. 74.

ball, w. If there is another such arrangement, of twice the size in every direction, the area of cross section of the rod would be four times as great, but the load to be carried would be eight times as great, and therefore the stress per square inch at a section would be twice as great—that is, the larger rod and ball would be twice as weak. As the stress would be twice as great and the length of the rod twice as great, the extension would be four times as great. The extension of the rod per foot in length would only be twice as great. In the same way a beam of cast iron, 1 inch square and 1 foot long, is 1,700 times too light to break with its own weight, whereas a beam of cast iron whose length, breadth, and depth are in the same proportion, if 1,700 feet long and 1,700 inches square in section, would break with its own weight. The deflection of similar beams similarly loaded is proportional to the square of their dimensions; but the deflection per foot of length is only proportional to their dimensions.

CHAPTER XIII.

GRAPHICAL STATICS.

133. The basis of all applications of mathematics in Physics and Engineering is the fact that any physical phenomenon which is **directional** (such as a force, a velocity, an acceleration, a stress, the flow of a fluid, &c.,) may be represented in a most perfect manner by a straight line. The general application of the geometry of the straight line to all physical calculations is a science called *Quaternions*, and it certainly is not my object to teach you any of the methods employed in this study to denote the relations of lines to one another. But with a box of drawing instruments and a sheet of paper, it is easy, by actually drawing the lines and measuring lengths, to solve many problems which would otherwise require a considerable knowledge of mathematics. This sort of graphical calculation having proved useful, it has attracted the attention of men who have leisure enough to make an elaborate study of its methods. It has unfortunately been dignified with the name of a new science. It has become a complicated weapon with which these men can attack all sorts of problems which are much more easily solved in other ways; and the result is that, instead of our having a few useful pages anywhere devoted to the subject, we have large treatises, adorned with numerous steel engravings, whose complications of lines frighten the student. It has been the same with many other useful processes; a few persons devote their attention to them and find that they are all-powerful, and every writer on the subject thinks it his duty to show how all sorts of problems are attacked. For instance, how few workmen know how to divide or multiply numbers or square or extract square roots by means of the slide rule! How few educated engineers, even, are able to make use of it! The owner of the instrument shop in which it is sold is seldom able to explain its use. It is known that very complicated problems can be worked out on it, but it is also known that to learn the working of these would require more time than it is worth while to devote to them, and, besides, they would be readily forgotten, whereas it is not sufficiently well known that the really useful processes of the slide rule can be taught to any man in a few minutes.

134. I shall begin this subject of Graphical Statics by giving some definitions, and indicating some propositions which can be proved by actual drawing. I shall speak of *forces*, because you have been accustomed to treat forces in the following way. I should have preferred merely to talk of *lines*, and to let these lines stand for any directional physical property.

135. Forces Acting at a Point.— The line A B (Fig. 74A) represents a force in *direction* by its own direction, in *amount* by its length to any scale we please, and in *sense* by its arrow-head, which shows that the action of the force is from A to B. It would not be correct to call this the force B A, because this is opposed to the sense of the arrow-head. The forces A O, O B, O C, O D (Fig. 75), all act upon a small body, O, or their lines of action when produced all pass through a point, O, in a large rigid body. The amount of each force is shown by the length of the line, representing it to some scale of so many pounds to the inch. Now to *add* these forces together in the most perfect manner—that is, to find a force called their *resultant*, which shall be exactly equiva-

Fig. 74A.

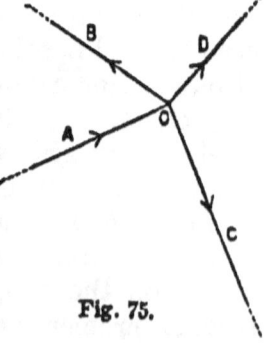

Fig. 75.

lent in its effects to all the above forces acting together—
we draw a polygon (Fig. 75A). Each side of this polygon
is parallel to and proportional to a force in Fig. 75 ; thus,
the side A corresponds to the force A O, and the arrow-
heads agree, and lastly the action indicated
by the arrow-heads is concurrent from A
round to D. Fig. 75A is always called **The
Force Polygon.** When it is *unclosed*, as it
is in the present case, we know that the
forces A O, &c. (Fig. 75), are not in equilibrium. To keep A O, O B, &c., in equilibrium a new force, called the *equilibrant* (see

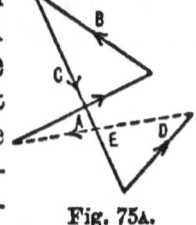

Fig. 75A.

GLOSSARY, Art. 218), must be introduced corresponding to
the side E (shown dotted), which will close the polygon, its
arrow-head being concurrent with the others. Now if we
want the resultant of A O, O B, O C, O D, it evidently acts
through O and corresponds to E, Fig. 75A, but with arrow-
head reversed. *The resultant of a number of forces is
equal and opposite to their equilibrant.*

Prove now the following exercises by actual
drawing :—

1st. The resultant of any number of forces does not
depend on the order in which they are drawn as sides of
the polygon.

2nd. Any lines or forces whatever which form a
closed polygon in any given order will form a closed
polygon if drawn in any other order.

3rd. In adding forces we may first find the resultant
of some of the forces, and then add together this resultant and all the rest of the forces. The answer will
always be the same, however we may group the forces
before adding them.

136. The Link Polygon.—We shall now consider
forces which do not necessarily act through one point.
Take, for example, the forces 1, 2, 3, 4 (Fig. 76). Draw
the unclosed force polygon 1', 2', 3', 4' (Fig. 77) with its
sides parallel to and proportional to the forces, and the
arrow-heads concurrent. Now the dotted line *b a*,

with its arrow non-concurrent with the rest, is parallel to and proportional to the resultant of all the given forces. But this does not tell us *where* the resultant force is situated, although it tells us its direction and amount.

From any point, o (Fig. 77), draw a line to the junction of 1' and 2' (it is easier to say draw the line o 1' 2'),

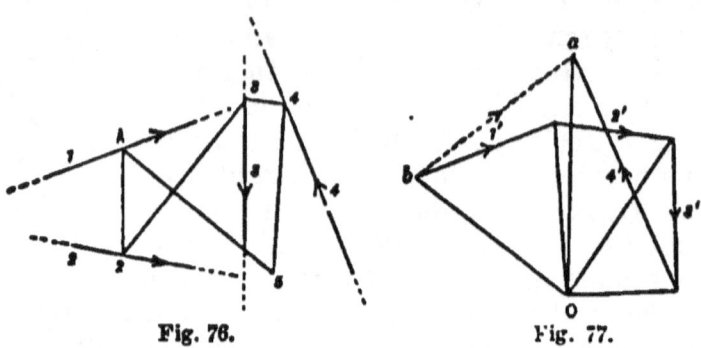

Fig. 76. Fig. 77.

o 2' 3', &c., to all the angles of the force polygon. Now construct a new unclosed polygon, with its corners on 1, 2, 3, 4 (Fig. 76), and its sides parallel to o 1' 2', o 2' 3', &c. (Fig. 77), its last side being parallel to oa, and its first parallel to ob. We have now found the point, 5 (Fig. 76), where the first and last sides of the link polygon meet. The resultant of the forces 1, 2, 3, 4, passes through this point, 5, and corresponds to the closing side, ba, in direction and magnitude. The new polygon is called THE LINK POLYGON of the forces relative to the *pole*, o. The position of A, the point at which we start to draw the link polygon, may be chosen anywhere, and hence there may be any number we please of link polygons for a given position of the pole, o. Again, there are any number we please of link polygons corresponding to any other position of o, and we can choose o where we please.

Suppose we find that when we are given the forces 1, 2, 3, 4 (Fig. 78), and we draw the force polygon (Fig. 79), and any link polygon (Fig. 78), that these are both closed, let us **prove that the forces are in equilibrium.**

A system of forces acting on a rigid body is not affected

Chap. XIII.] FORCE AND LINK POLYGONS. 139

by introducing any number of forces which separately balance one another. Now let a force represented by the length of the line o 1' 2' act at the point A in the direction B A, its sense being shown by the arrow-head near A, and let an equal force act at B in the direction A B, its sense being opposite to that of the force at A. These two forces are in equilibrium with one another, and they cannot therefore affect the

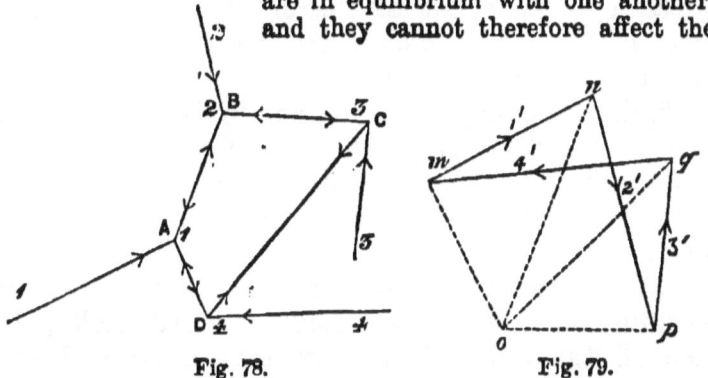

Fig. 78. Fig. 79.

original system of forces in any way. Similarly, the forces shown by the arrow-heads in B C, C D, D A are introduced, every pair balancing one another.

Now we see that the three forces at the point A are in equilibrium with one another, because they are parallel to and proportional in amount to the sides of the triangle o m n (Fig. 79), and corresponding arrow-heads would run right round the triangle. Similarly, there is equilibrium at every other corner of the link polygon A B C D; hence all the forces are in equilibrium, and hence the forces, 1, 2, 3, 4, taken by themselves, must be in equilibrium.

The theorems which we wish students to prove by construction can be proved to be generally true, reasoning from the fact that a number of forces acting at a point can only have one resultant.

137. We see, then, that the force polygon alone is sufficient to find the resultant of any number of forces if the forces meet at a point, but we need also the link polygon if the forces do not meet at a point.

The *link* polygon really shows that the sum of the turning moments of the forces 1, 2, 3, 4 (Fig 76) about any point is equal to the moment of the resultant about the same point. The *force* polygon pays no regard to turning moments of forces; it merely tells us about the resultant of the forces, supposing that they all passed through the same point.

138. **You ought to prove the truth of the following statements by actual drawing :—**

1st. The direction of the resultant of any number of forces is independent of the order in which we draw them in the force polygon and draw between them the sides of the link polygon.

2nd. In adding forces we may first find the resultant of some of the forces, and then add together this resultant and all the other forces. The result will always be the same, however we may group the forces before adding them.

3rd. If the force polygon of a number of forces is closed, and if we can draw a closed link polygon, then all the link polygons we may draw will also be closed.

4th. If any other pole be taken in Fig. 77, and another link polygon be drawn and a new point 5 (Fig. 76) is found, both of the points so found lie in a straight line parallel to $b\,a$ of Fig. 77.

> You will also find, and it is easy to prove, that the locus of the point in which any two sides of the link polygon meet is parallel to the line which closes the corresponding portion of the force polygon. Again, if b is taken as pole instead of o, the last side of the link polygon is found to be in the direction of the resultant of the forces 1, 2, 3, 4; and, generally, any side of the link polygon is in the direction of the resultant of the corresponding number of the given forces. Thus, if b is taken as pole, 4 5 (Fig. 76) becomes the resultant of the forces 1, 2, 3, 4, and 3 4 becomes the resultant of the forces 1, 2, 3. It is evident from this that the direction of the resultant of any two lines or of any number of lines which meet at a point passes through their point of intersection.
>
> A system of forces may not reduce to a resultant force, but be equivalent to a *couple*. When this is the case the force polygon is closed, and the first and last sides of any link polygon that may be drawn are parallel to one another. You may also find it worth your while to prove by construction this statement; If two link polygons are drawn for two

Chap. XIII.] EXAMPLES IN GRAPHICAL STATICS. 141

positions of the pole o, the corresponding sides of the two polygons meet in points which lie in a straight line parallel to the line joining the two positions of the pole o.

139. If you have been able to make a few drawings such as I have been speaking about, and so take an interest in this easy and instructive method of working mechanical exercises, you ought to work by means of it a few such exercises as the following :—In Fig. 80 A B represents a ladder whose centre of gravity is at G, and weight 300 lbs. A string fastened to it at C in the direction C O keeps it in equilibrium, its end A resting on the smooth wall O A, and its end B on the smooth floor O B. Find the pull in the string and the reactions at A and B.

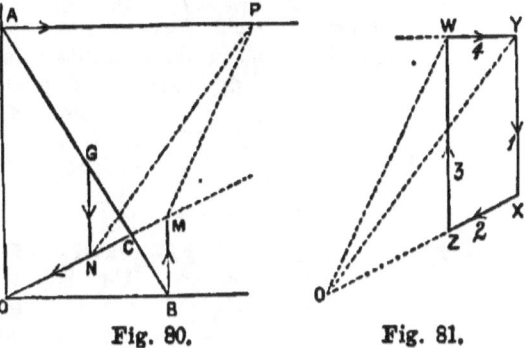

Fig. 80.   Fig. 81.

The forces acting on the ladder are shown by the arrows. Draw Y X (Fig. 81) vertically to represent the weight of the ladder. Draw X O parallel to O C, and take O anywhere in this line. Use O as pole of the force polygon. Join O Y. Now the link polygon is M N P M, and drawing O W parallel to P M, and W Z parallel to B M, we find that Y X Z W is the force polygon. The lengths of X Z, Z W, W Y represent the forces at C, at B, and at A.

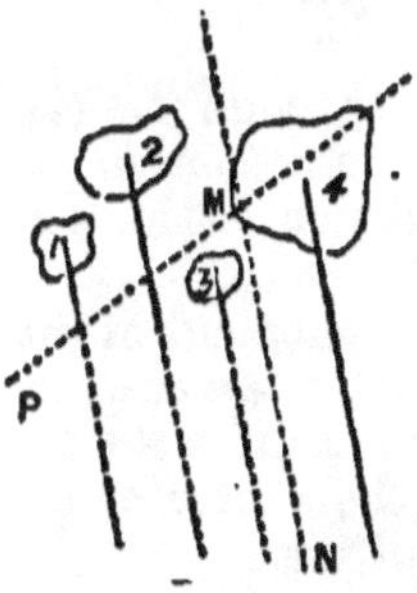

140. Again, the centre of gravity of a number of masses or a number of areas may be found easily by this method. Thus, let there be masses or areas, m_1, m_2, &c., whose centres are at the points 1, 2, &c. (Fig. 82). Draw the parallel forces 11, 22, &c., in any direction proportional to $m_1\ m_2$, &c., and find the resultant by the above method. Suppose M N to be the direction of the resultant. Now repeat the process, taking the parallel forces of the same magnitudes as before, but in a different direction, and let M P be the direction of the resultant. Evidently M, where these lines meet, is the

Fig. 82.

centre of gravity of the masses or areas. This method may often prove useful, for areas especially. Thus, to find the centre of gravity of any given area, divide it into any suitable number of parts, so that the centre of gravity and area of each part may be found easily. If we divide the area by parallel lines, these lines may be drawn equidistant, and the area of each part is approximately given by the length of the line which separates it from either of its neighbours. A repetition of the process has been employed to determine the moment of inertia of the area about any given line.

141. I do not advise students to adopt this link polygon method of finding centres of gravity or of calculating moment of inertia. A practical engineer will always apply the ordinary formula to find the centre of gravity of an area. Thus, if you want the centre of gravity of the figure M N O P (Fig. 83), draw two parallel lines, G H, K O, touching the figure at two opposite sides. Draw a line, K G, at right angles to G H, and divide it into any number of equal parts, each equal to d. Draw the lines A B, C D, &c., parallel to G H, so that they are at the distance d apart, the distance from A B to G H, or from Y Z to K O being $\tfrac{1}{2}d$. It is evident that if N X P is a line parallel to G H through the centre of gravity, then approximately

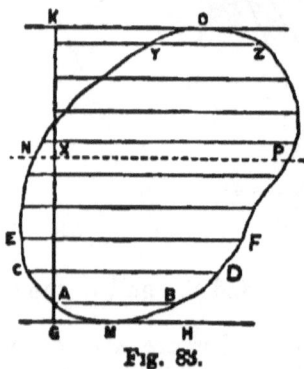

Fig. 83.

$$G X = \tfrac{1}{2} d \cdot \frac{AB + 3CD + 5EF + \&c.}{AB + CD + EF + \&c.}$$

We have thus obtained one line through the centre of gravity, and in a similar way we may obtain another such line, and their point of intersection is the centre of gravity required.

In the same way we may obtain the **moment of inertia** I of any area about any line; or, as is often the case, suppose we wish to find the moment of inertia of M N O P about N X P, a line which passes through the centre of gravity. Evidently the moment of inertia about G H is

$$I = \frac{d^3}{4} (AB + 9 CD + 25 EF + \&c.).$$

Now, it is well known that *the moment of inertia of an area about any line is equal to its moment of inertia about a parallel line through its centre of gravity, together with the product of the area into the square of the distance between the two lines.* Hence, the moment of inertia about N × P is

$$I_0 = I - G X^2 \cdot d \, (AB + CD + EF + \&c.).$$

It will be found in practice that this easy way of carrying out simple ideas is better than the complicated use of the link polygon method for finding moment of inertia. If the area may be divided into rectangles whose sides are parallel to and perpendicular to the axis, we need not subdivide these rectangles. It must be remembered that the moment of inertia of a rectangle about any axis is equal to the area of the rectangle multiplied by the square of the distance of its centre of gravity from this axis, plus the moment of inertia of the rectangle about a parallel line through its centre of gravity. Thus, the rectangle A B C D (Fig. 84) is known to have about N M N the moment of inertia $\frac{AB \cdot BC^3}{12}$, so that the moment of inertia of the rectangle about O' O O' is

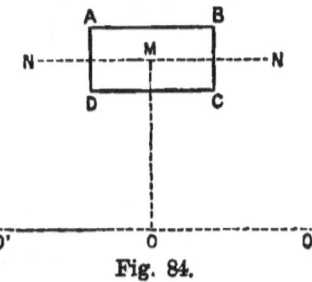

Fig. 84.

$$\frac{AB \cdot BC^3}{12} + AB \cdot BC \cdot MO^2, \text{ or } AB \cdot BC \left(\frac{BC^2}{12} + MO^2 \right).$$

The student will find it good exercise to take a few sections of angle-iron, T-iron, rails, and other specimens of rolled iron, and find the position of centre of gravity of each section, and the moment of inertia of each area about any line through the centre of gravity. The exact forms ought to be taken from real specimens. If the area is symmetrical, one line through the centre of gravity can always be found by mere inspection.

142. When the moments of inertia of an area about any three axes through a point are known, the moment of inertia about any other axis through the same point may be found; because if a distance be measured from the point along an axis which is equal to the reciprocal of the

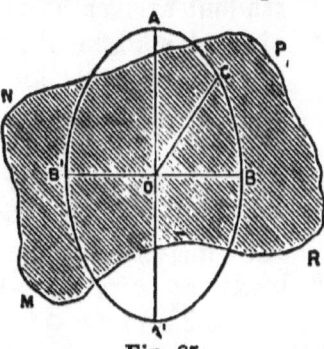

Fig. 85.

radius of gyration of the area about the axis, the extremities of all such measured distances lie in an ellipse. The *principal axes* of the area are in the directions of the major and minor axes of this ellipse. Thus, if for any area, M N P R (Fig. 85), the least moment of inertia is about an axis, O A, and is $\frac{m}{OA^2}$, and if the greatest moment of inertia is about O B, and is $\frac{m}{OB^2}$, then A B A' B' being an ellipse whose major and minor axes are A A' and B B', the moment of inertia about an axis, O C, is $\frac{m}{OC^2}$. This theorem of Poinsot's is proved in all elementary treatises on dynamics. The student will find it useful to prove it by actually finding the moments of inertia of any area about a number of axes.

CHAPTER XIV.

EXAMPLES IN GRAPHICAL STATICS.

143. Diagrams of Bending Moment.—Let A B (Fig. 86) represent the length of a beam which has three vertical loads—1, 2, 3. To find the vertical supporting forces at A and B, draw the unclosed force polygon, K L (Fig. 87)—before the student arrives at this part of the book he will probably have drawn other force polygons where all the sides were really in the same straight line—1, 2, and 3 (Fig. 87), representing in direction and magnitude the three loads of Fig. 86. Choose any point, O. Join O K, O 1 2, O 2 3, and O L. Now draw the link polygon (Fig. 86), beginning at any

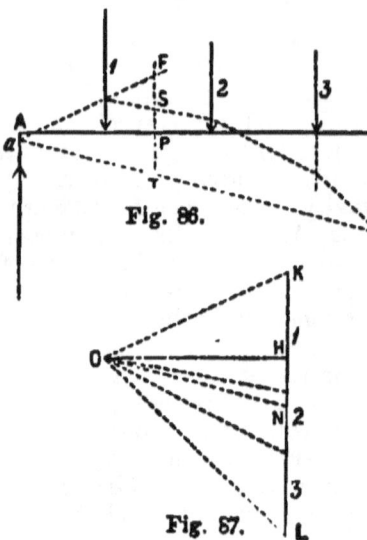

Fig. 86.

Fig. 87.

point, A, in the vertical from A, and ending in the point b. Now ab is the side wanting in the force polygon. Draw ON (Fig. 87) parallel to ab (Fig. 86). Then LN is the amount of the supporting force at B, and NK is the amount of the supporting force at A. Also, draw any vertical line, ST (Fig. 86). Then *the length* ST, *intercepted by the sides of the force polygon, represents the bending moment of the beam at any point*, P, on some scale which it is easy to find.

To prove this. Draw OH horizontally. The moment at any point, P, due to the supporting force, NK at A, is NK × AP; and this is equal to OH × FT, for, by similar triangles,

$$AT : TF :: ON : NK,$$

and therefore

$$AP : TF :: OH : NK.$$

This second proportion gives NK.AP equal to OH.TF.

In the same way the moment at P, due to the force 1, is OH.FS; and hence the true moment at P, being the difference of these, is OH.ST. We see, then, that if, for example, the beam is drawn to a scale of

1 foot, represented by x inches,

and if the loads are drawn to a scale of

1 pound, represented by y inches,

then ST is the bending moment at P, on a scale such that

1 pound-foot is represented by $\frac{xy}{OH}$ inches.

If the load is not concentrated at a number of points, it is usual to imagine it divided into a number of loads, each of which acts at one point. The diagram of bending moment is drawn in the way which I have just described, and then for the polygon with its straight sides we substitute a curve which touches all the sides of the polygon.

After you have found a diagram of bending moment, if you wish to see the effect of additional loads, draw a diagram for these loads as if they acted alone, but take care that the horizontal distance, OH, is the same as before. Add together the ordinates of your two diagrams to get your new diagram of bending moment for all the loads.

144. Shape of a Loaded Beam.—When we know the bending moment at every cross section of a beam, it is easy to draw its shape, the vertical dimensions being increased in any scale we please. Divide the bending moment at every place by the modulus of elasticity and by the moment of inertia of the cross section there, and draw a diagram which shows at every place the value of $M \div EI$. Regard this as a diagram which shows the amount of a new kind of continuous load per foot in length of the beam. Draw a new diagram as the diagram of bending moment was drawn. This shows at every place what has been the amount of vertical yielding of the beam. If the scale of EI is diminished, the scale of the deflection of the beam is increased in the same proportion. The proof of this proposition cannot easily be given without the use of mathematical expressions, which I prefer to keep out of this book.

145. Hinged Structures.—Before giving you the easy rules by which we find the forces acting in structures built up of a number of tie-bars and struts hinged together, such as roof-principals and iron girders, I think it advisable to mention a few facts about the geometry of certain figures which consist of points joined by straight lines. . They are such that;—

First, we are able to resolve them into a number of closed polygons. *Second*, each side in a figure is a side of two polygons, and only two. *Third*, from each point there come at least three straight lines. *Fourth*, each line passes

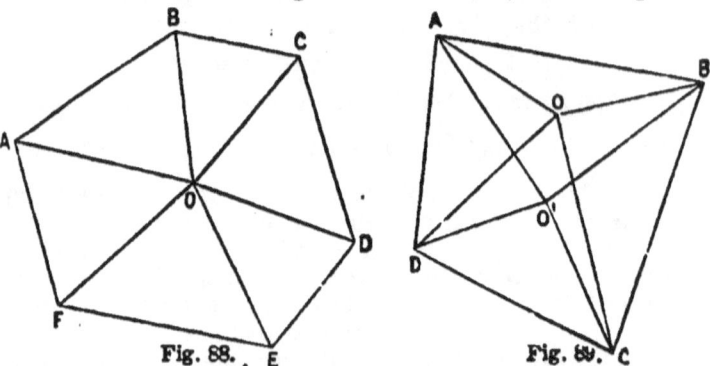

Fig. 88. Fig. 89.

through at least two points. If a line passes through three points or more we must consider its segments as distinct sides. Thus, O A B C D E F (Fig. 88) is a figure which satisfies the conditions; it consists of seven points, joined by twelve lines, and it contains seven closed polygons or triangles. Also A B C D O O' (Fig. 89) is a figure consisting of six points joined by twelve lines; to satisfy condition (2) given above, we must not consider A B C D as one of the polygons, we only take into account the eight triangles.

It will be found that in all figures satisfying the above conditions the number of closed polygons plus the number of points equals the number of sides plus two. This is proved by taking any suitable figure and adding a new side; it is found that the sum of the number of closed polygons and the number of points is also increased by unity. This is, indeed, the relation between the number of faces, summits, and edges of a polyhedron, and all the figures of which we speak may be regarded as projections of polyhedra.

145a. Straight-line figures generally may be divided into:—1. Deformable figures, or those which may alter in shape, the lines retaining their original lengths. 2. Figures perfectly stiff. 3. Figures which would be perfectly stiff, even if we removed one or more lines. It may be shown that a figure belongs to class 1, 2, or 3, according as the number of its sides is less than, equal to, or greater than, double the number of points minus three. It is evident that in a figure of the third class the lengths of the extra lines may be expressed mathematically in terms of the other sides. Thus, if there are a points, b sides, and e polygons in any figure, we generally find that there exist $b - 2a + 3$ necessary conditions regarding the lengths of the sides, and if these conditions are not satisfied the figure has no existence.

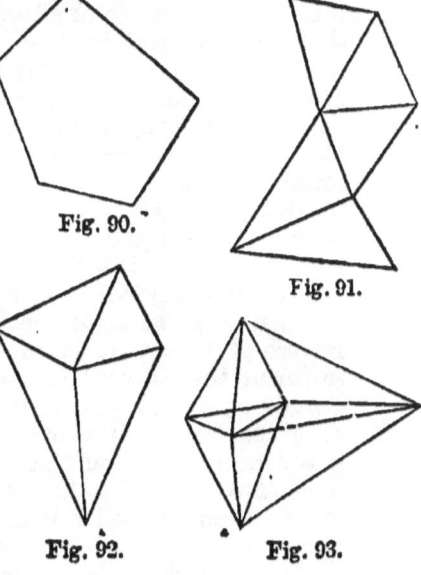

Fig. 90.

Fig. 91.

Fig. 92.

Fig. 93.

If in a given figure $b - 2a + 3$ is negative, then $2a - b - 3$ further conditions must be given, to add lines to the figure before it can become stiff.

Thus, Fig. 90 is deformable (class 1); Fig. 91 is stiff (class 2); Fig. 92 contains one extra side (class 3); Fig. 93 contains four extra sides (class 3).

146. In every problem in Graphical Statics it will be found that we are concerned with two figures which have a certain reciprocal relation to one another. We shall now define this relation.

Two figures are said to be **reciprocal** to one another when to each line in one figure there corresponds one parallel line in the other figure, and the lines which meet in a point in one figure correspond to the lines forming a closed polygon in the other figure. Thus Fig. 74 and Fig. 75 are reciprocal, and Fig. 78 and Fig. 79 are reciprocal.

It may be shown that unless a figure satisfies the conditions given in Section 145 it cannot have a reciprocal figure, but that all figures which satisfy those conditions do admit of reciprocal figures. Thus Figs. 88 and 89 admit of reciprocal figures. So also that a figure composed of any number of closed polygons, A B C D E, A' B' C' D' E', &c., with the same number of sides, A A', B B', &c., being joined, shall admit of a reciprocal figure, it is necessary and sufficient that the points in which all agreeing sides meet, if produced, shall lie in a straight line. Thus, the points where A B and C D meet, A' B' and C' D', A" B" and C" D", etc., all lie in one straight line, and so for the other sides. We observe, therefore, that any number of link polygons obtained from the same forces may form a figure which admits of a reciprocal figure.

Generally, we may say;—that where any figure composed of points joined by straight lines shall admit of a reciprocal figure, it is necessary and sufficient that it is the projection of a polyhedron with plane faces.

That a given figure shall admit of one, and only one, reciprocal figure, it is in general necessary and it is sufficient that it contains one extra line. If it does not contain an extra line, then it is necessary for it to satisfy one condition, if it is to have one and only one reciprocal figure. If it is deformable, it only admits of a reciprocal figure when it satisfies as many conditions plus one as there are new lines to be traced to render it stiff. If it contains two or more extra lines, it admits of any number of reciprocal figures.

In fact, a figure admits in general of none, or of one, or of any number of reciprocal figures according as the number of its points is greater than, equal to, or less than, the number of its closed polygons.

There are exceptions to these rules, which I shall not enter into.

147. In any structure, such as the principal of a roof and many girders of bridges, formed of many different bars, if we assume that each bar is subjected to direct pull or direct push—that is, if we assume that the forces with which a bar may act at the two joints at its ends are in the line joining those two joints—then it is easy to calculate the direct push or pull which each bar has to resist when the structure is loaded. This is assuming that the external loads are all applied at the joints of the structure, and that these joints are really the centres of frictionless hinges.

In actual practice, however, the joints are usually stiff—that is, the bars are really subjected to bending and shearing stresses as well as to direct compressive or tensile stress. But it is found that the strength of many structures, when tested, is approximately the same as if they were hinged structures.

The conditions which enable us to calculate the stresses in a hinged structure are:—

1. All the external forces are in equilibrium with one another.

2. The pulls and thrusts and loads acting at any one joint form a system of forces which are in equilibrium with one another.

3. A piece connecting two joints pulls or pushes one of the joints with the same force with which it pulls or pushes the other.

Having determined the amount of pull or push in each bar, we can find the most suitable cross section to give the necessary strength, if we know the material, by using the results of Chap. XI. To illustrate what we mean by a joint, find the stresses in the two pieces, O A and O B (Fig. 94), which have a hinged joint at O, when there is a vertical load of 2,000 lbs. at O. The piece O A acts upon O with a certain force, and so does O B, and the three forces balance. Draw a triangle (Fig. 95), with its sides parallel to O C, O B, and O A, in such a way that the side *c* represents on some scale the force of 2,000 lbs. The direction of O C, being a pull downwards, shows the direction in which we must draw the arrows concurrent round the triangle *c b a;* then *b* represents to the same scale the force in O B, and the arrow shows its action on O; thus we see O B pushes O, so we call O B a *strut*. We also know that at the other joint of O B, say B, the piece O B exerts a pushing force of amount *b*. Similarly, A O is a *tie*—that is, it exerts a pull upon O of the amount *a*.

148. Let us now consider the **roof-principal** shown in Fig. 96. Certain loads are given acting at the joints,

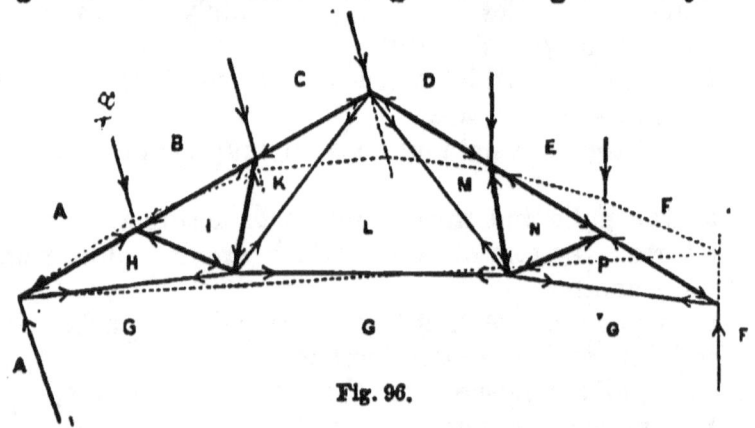

Fig. 96.

and we know that the structure is supported by two forces or reactions at its two ends. Our first step is to find these two supporting forces. They must be in equilibrium with all the external loads.

Now, it is well known that we must be given either the *direction* or the *amount* of one of these supporting forces, else the problem becomes indeterminate. It is usual to be told that one or other supporting force is vertical. This condition is arrived at in practice by having at one end of a structure a little carriage with wheels resting on a horizontal plate of iron.

The notation which we use is due to Mr. Bow. It

very materially simplifies the process of calculation. You observe that every *space* between two forces in Fig. 96 is indicated by a letter. The line which separates the space A from the space B is called A B, and corresponds with the line A B in Fig. 97. A point is indicated by the letters of the spaces which meet at that point. Thus, A G H is the end of the roof-principal.

Suppose the supporting force at the point F G P is known to be vertical. We must first find the amount of

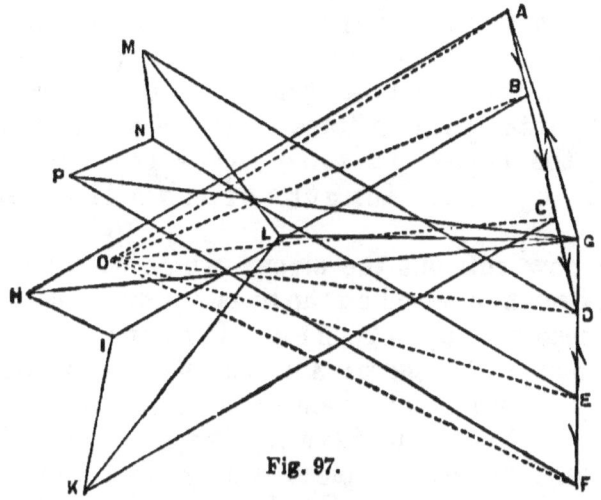

Fig. 97.

the force F G, and the direction and amount of the force A G.

Draw the force polygon, A B C D E F, Fig. 97. We see that to close it we need two lines to join F and A. Now, one of these, F G, is vertical. Take O as pole. Join O A, O B, O C, O D, O E, O F in the usual way. Draw the link polygon, shown dotted in Fig. 96, commencing at A G H. Now, O G (Fig. 97) is parallel to the last side of it, and thus we find F G and G A, the supporting forces at the end of the principal. Having found the two supporting forces, F G and G A, we proceed as follows :—We have the closed force polygon, A B C D E F G A. The arrow-heads shown on this force polygon are not to be rubbed out during the calculation, and in practice we mark them in ink. All other arrow-heads which we draw on Fig. 97 may require to be rubbed out, and ought only to be marked in pencil. We must begin our calculation at a joint where only two pieces meet, and where one force which acts there is given. Now at the joint A G H we know the force A G. In Fig. 97 draw A H and G H parallel to the pieces A H and G H of Fig. 96. Put arrows on the sides of the triangle G A H concurrent with the arrow on G A. Now we see by the arrows that the piece A H *pushes* the joint with a force represented to scale by the length of the line A H (Fig. 97). We know, then, that A H is a *strut*, since it pushes, and we know the total pushing force in it. Similarly, H G is a *tie*, and the *total pulling force* in it is represented by the length of the line H G in Fig. 97.

We now rub out the arrows which we are supposed to have drawn in pencil on the lines A H and H G (Fig. 97), and proceed to the joint A B I H. It must be remembered that although the pieces A H and B I are in the same straight line, we regard them as two separate pieces.

Now we know the force A B, we also know that the force with which the piece A H pushes the joint is represented by the length of the line A H (Fig. 97). Draw, then, H I and B I (Fig. 97) parallel to the pieces H I and B I (Fig. 96). We have thus a polygon, A B I H. The force A B (Fig. 97) tells us how to pencil arrow-heads concurrently round this polygon. When we do this we find that the piece B I pushes the joint with a force represented by the line B I (Fig. 97), so that B I is a strut. Also I H is a strut, in which the stress is represented by the line I H (Fig. 97). We proceed in this way from joint to joint, always taking care to rub out our pencilled arrow-heads when we proceed from one joint to the next. The *lengths* of the lines in Fig. 97 give the *magnitudes* of the forces in the pieces of the structure. It is easy to prove that, if no mistake is made, no discrepancy will appear when the drawing is being finished—that is, when we are returning to the joint with which we began.

If in Fig. 97 the points K and I are found to coincide, this evidently means that the piece K I is unnecessary in the structure. If, again, we find that we cannot close one of our little polygons in Fig. 97, we ought to proceed to new joints, and, possibly, when we again consider the joint with which we had difficulty, we shall be able to close its polygon. If we still find difficulty, it must be caused by two or more joints, and the pieces connecting these are evidently unnecessary to the structure. If we find in Fig. 97 two points with the same letter, we evidently require to add a new piece to the structure, which will exert a force equal to the distance between these two points.

No explanation in writing will enable the student to master this beautiful **method of determining the stresses in structures.** He must select structures, apply loads to the joints, and calculate the various stresses for himself. When he has made four such calculations, he will know nearly all that can be said on the subject.

149. Roofs.—It is not my object here to describe the construction of a roof or a bridge. For such information the student must examine real structures for himself; he must read Tredgold's treatise on roofs, and examine many good drawings of roofs and bridges.

Suppose, for instance, that he finds a roof, somewhat like his own, to weigh—including possible snow, &c.— 20 lbs. per square foot of horizontal area covered. Suppose his principals are to be placed 8 feet apart, the span being 50 feet, then each principal has to support about

$$8 \times 50 \times 20, \text{ or } 8,000 \text{ lbs.}$$

Now, if Fig. 98 is the shape of his principal, as A B, B C, C D, and D E are all equal, we may suppose that, however the roof covering may be supported by the principal, the piece of rafter, A B, or any other of the divisions, supports 2,000 lbs. The joint B gets half the load on A B and half the load on B C; consequently, the load at the joint B is taken to be 2,000 lbs., and similarly for C and D. The joints A and E do not need to get loads, because we have afterwards to calculate the total forces at A and E by the link-polygon method.

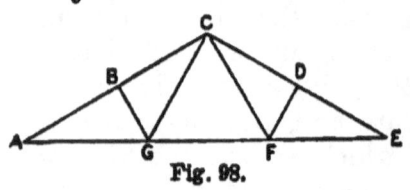

Fig. 98.

When the above vertical loads have been given to the joints, we have to consider wind pressure on one side of the roof. If we suppose, as we reasonably may, that 40 lbs. per square foot is the greatest pressure of wind ever likely to occur on a surface at right angles to the direction of the wind, then the normal pressure per square foot on roofs of the following inclinations may be taken from the following table, which is obtained from Hutton's experiments.

TABLE X.
Normal Pressure of Wind against Roofs.

Angle of Roof.	Normal Pressure.	Angle of Roof.	Normal Pressure.
5°	5·0	50°	38·1
10°	9·7	60°	40·0
20°	18·1	70°	40·0
30°	26·4	80°	40·0
40°	33·3	90°	40·0

Thus, if the portion of one slant side of the roof between two principals has an area of 240 square feet, and if the inclination of the roof is 30°, say, then 240 × 26·4, or 6,336 lbs., has to be supported by each bay. Transferring this to the joints, we see that at B (Fig. 99) we have the vertical load x B, or 2,000 lbs., due to weight of roof, snow, &c., and also Y B, or one half of 6,336 lbs., normal to the roof, and due to wind. Complete the parallelogram, and evidently Z B is the load at the joint B which we must use in our calculations.

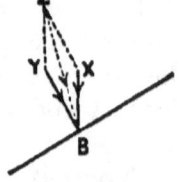

Fig. 99.

The student will find that if a roof-principal can only be supported by a vertical force at a certain end, the stresses in the structure are greatest when the other side of the roof is acted on by the wind.

I would advise every student to design at least one

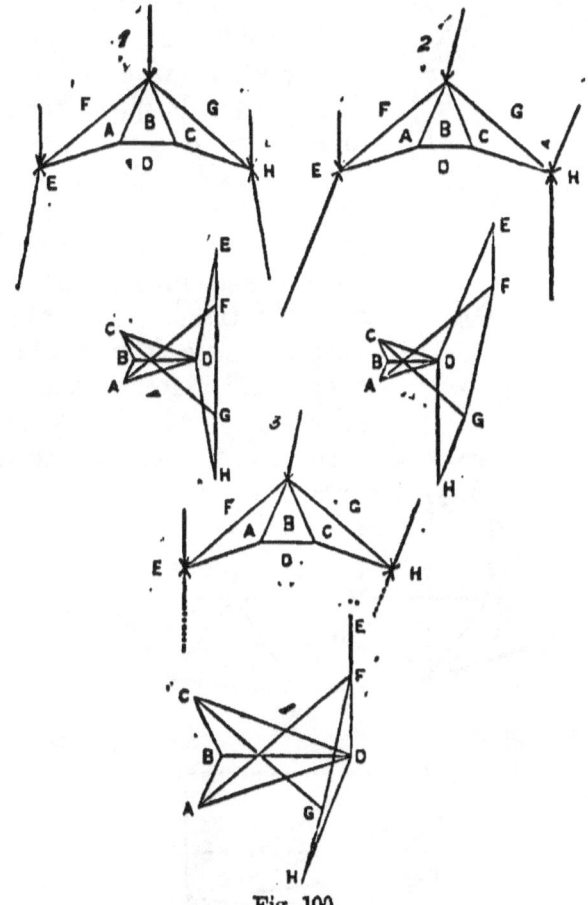

Fig. 100.

timber and one iron roof, making detail drawings of all the joints, etc., referring much to books and drawings, and writing out complete specifications. The following **examples of stress diagrams** (Figs. 100, 101, 102) may be useful for reference.

150. Every joint in a real structure is usually a stiff joint; so that every piece may really be subjected to bending, as well as to direct compressive and tensile stresses. A general method of taking stiffness of joints into account is quite unknown; but as we have discussed *bending* we see pretty clearly what is the effect of a stiff joint, and in some cases we are able to make calculations on the subject. It may generally be assumed that the strength of a structure is greater if the joints are stiff than if they are merely hinges. This is not always the case, and, from the indeterminateness of the problem of finding the stresses in a structure whose joints are stiff, many large bridge trusses are at present made with

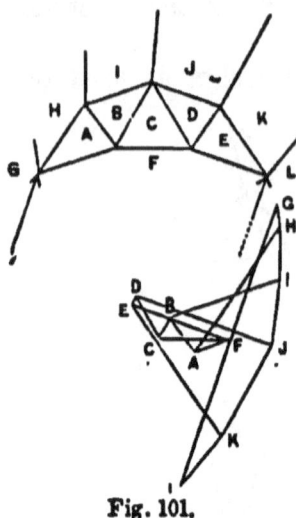

Fig. 101.

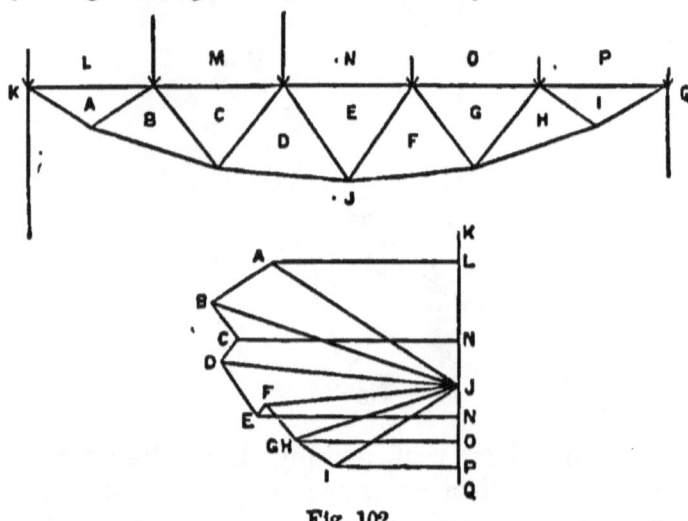

Fig. 102.

nearly all their joints hinged. In roof-principals the joints are made stiff, rather for the purpose of *stiffening* the whole structure—that is, increasing its resistance to change of shape—than for the sake of strength. In a roof all joints of struts are usually made stiff. What we shall now say is of more importance in bridges than in roofs.

If two or more pieces of a structure are in a straight line with one another at joints where they meet, it is usual for strength to make the joints between them quite rigid. Thus the pieces A H and B I of Fig. 96, or A B and B C of Fig. 98, ought to form one bar. But this is only useful when the pieces in question are struts, and our reason for the continuity of the pieces is that a strut is stronger when its ends are fixed than when its ends are not fixed. Thus the piece B I (Fig. 96), will resist a greater thrust if it is continuous with A H and C K than if it were hinged with these pieces. (See Art. 129.) It is not good in all cases to fix the end of a strut by rivets, &c., instead of a hinge; because the benefit due to fixing an end may be more than counter-balanced by the evil effects of bending introduced to the strut through the joints by a tendency to change the angle which the strut makes with the piece to which it is fixed. The common sense of the engineer will always enable him to decide as to the judiciousness of fixing the end of a strut.

151. Sections of Structures.—It is often of considerable importance to find immediately the stresses in pieces of a structure which are not near the ends. If we can draw any surface which will cut through the pieces in question, we can calculate the stresses in these pieces directly, supposing the pieces are only three in number. Thus, the section A C E (Fig. 103) cuts the pieces B A, B C,

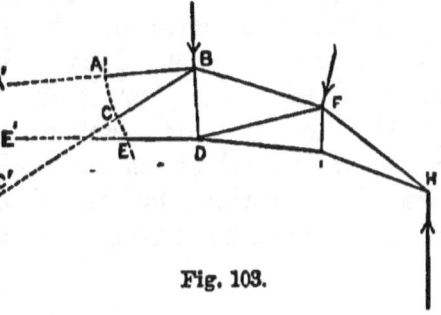

Fig. 103.

and D E. Now not only the whole structure, but every part of it is kept in equilibrium. What forces keep the part A H E in equilibrium ? They are the known forces at B, F, and H, together with three unknown forces whose directions are B A A', B C C', and D E E'. Given the directions of three forces which equilibrate a number of known forces, we know that they may be determined in magnitude by the link-polygon method. Sometimes the link-polygon method is more troublesome than the following:—To find the push or pull in D E E'. We know (Art. 23) that the moment of the force in E E' about the point B is equal to the sum of the moments about B of all the external forces (for the forces in the directions A A' and C C' have no moment about B, since they pass through it). Let the algebraic sum of the moments of the external forces be actually calculated, multiplying numerically each force by its perpendicular distance from B. This sum, divided by the perpendicular distance from B to E E', will give the force in E E'. If the algebraic sum gives a moment tending to turn the structure about B against the direction of the hands of a watch, the force in E E' is a pulling force acting from E towards E', and therefore the piece D E is a tie.

It will be observed that if we wish to know the stress at any section of any loaded structure, we must consider that the parts of the structure on any one side of this section are in equilibrium. Thus, if A and B are the two parts of the structure, consider the equilibrium, say, of B. Now, B is kept in equilibrium by the external forces or loads which act on B, and by the forces which act on B at the section. Of course it is A which causes these forces to act on B through the section; but in calculations concerning them we do not need to consider A or the loads on A.

CHAPTER XV.

SUSPENSION BRIDGES, ARCHES, AND BUTTRESSES.

152. Loaded Links.—Let A C, C D, D E, and E B be four links hinged together at C, D, and E, and supported somehow by hinges at A and B, and, neglecting the weights of the links themselves, let x, y, and z be forces acting at the three joints, so as to make the links take the positions shown in Fig. 104. Take any point, o (Fig. 105), and draw lines $o\,m$, $o\,n$, $o\,p$, and $o\,q$ parallel to the links, and from any point, m, in $o\,m$, draw $m\,n$ parallel to the force z, $n\,p$ parallel to the force y, and $p\,q$ parallel to the force x; then it is easy to prove that the lengths of the lines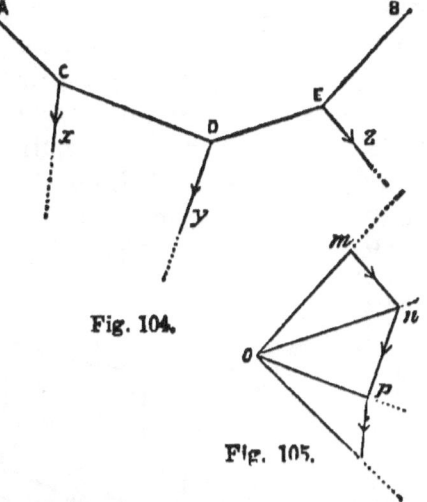

Fig. 104.

Fig. 105.

$m\,n$, $n\,p$, and $p\,q$ are proportional to the forces z, y, and x, and the tensile forces in the links are proportional to the lengths of the lines $o\,m$, $o\,n$, $o\,p$, and $o\,q$. For it is evident that the three forces at E, keeping the joint in equilibrium as they do, must be proportional to the sides of the triangle $o\,m\,n$. If you put arrowheads on $o\,m$ and $o\,n$ concurrent with the one already on $m\,n$, you will see that the bars B E and D E do not push the joint E, they pull it and are tie-bars. Thus, then, the lengths of the lines in Fig. 105 represent to some scale all the forces acting at the joints C, D, and E.

153. Loaded Chain.—If, now, we want to find the pull in every part of one chain of a suspension bridge, and to draw the shape of the chain, it is first necessary to know the weight of the bridge at every place. This weight is probably supported by two chains, so, as we have only one chain to deal with, we only take half the weight of the bridge. We will suppose that there is no long girder or other support for the bridge but the chain. It is usual to suspend the supporting beams of the roadway from the chain by vertical iron rods, placed at equal horizontal distances from one another. We may imagine the roadway to be as heavy at one place as another, so that the pull in all the rods will be the same. Suppose there are ten rods, and in each a pull of 20 tons. Draw ten equidistant vertical lines (Fig. 106) to represent the rods. We must get another condition before we can draw the chain. Let it be this, that the chain in the middle where it is horizontal shall be capable of withstanding a pull of 200 tons. Now draw o ii horizontally (Fig. 106A), and make its length on any scale represent 200 tons. Make ii A and ii B on the same scale represent each 100 tons (if your chain is to be symmetrical), and

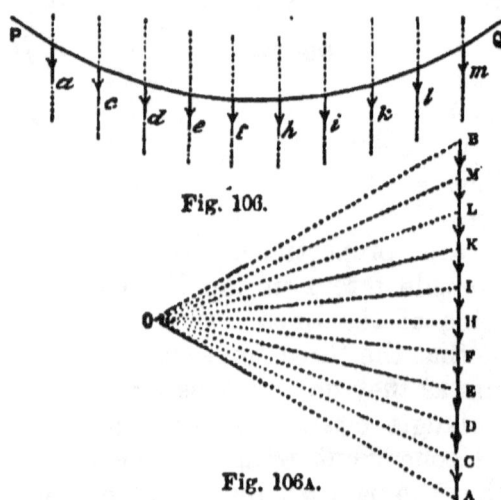

Fig. 106.

Fig. 106A.

divide them up so that each portion represents 20 tons— that is, the vertical load communicated to the chain by each tie-rod. Now join o with each point of division in A B. Suppose now that P (Fig. 106) is one point of support of the chain, draw P a (Fig. 106) parallel to O A

(Fig. 106A), ac parallel to oc, cd parallel to oD, and so on till you reach the point Q, which I suppose to be on the same level as P. Of course, the points of support, P and Q, may be anywhere on the lines a P and m Q. It is quite evident from what you have already learnt that the pull in any part of the chain is represented by the length of the line from o, which is parallel to it in Fig. 106A, and it is also evident that the chain will take this shape without any tendency to alter.

154. We began by assuming a pull of 200 tons in the part fh, where the chain is horizontal. We might have assumed a pull of 300 tons in fh; this would have caused the chain to hang in a flatter curve. Assuming a pull of 100 tons in fh, we should have obtained a greater difference of level between P and h.

It will be found that in the present case, where the load is supposed to be uniformly distributed along the horizontal, the links would just circumscribe the curve called a parabola. With any other distribution of load they will fit some other curve than a parabola, but in any case you know now how to draw the shape of such a chain, and to determine the pull in any part of it.

155. Arched Rib.—If instead of a hanging chain you wanted to use a thin arched rib to support your roadway, then if you have numerous vertical rods by which to hang your load to the rib, and if the distribution of the load is known, you can draw the curve of the rib in exactly the same way, but it will now be convex upwards of course. With uniform horizontal distribution of your load you will get a parabolic rib. The difference between the two cases is this: a slight inequality in your loads or a temporary alteration will only cause the chain to take a slightly different position for the time, and it will get back to its old shape when the old loading is returned to; whereas the arch is in a state of unstable equilibrium, and as it is very thin, so that it cannot resist any bending, a slight change of loading will very

L

materially alter its shape and it will get destroyed. Such a rib or series of struts is either stayed with numerous diagonal pieces or else it is made very massive, so that should the line like P h Q (inverted), which is supposed to pass everywhere along its axis, deviate a little from this position, the rib may resist alteration of shape by refusing to bend.

156. The load carried by an arch may either be hung from it by means of tie-rods, or else it may rest on the top of the arch, the weight being carried from the different parts by means of struts or pillars of iron, stone, or brick, or the arch may be levelled up to the roadway by means of a solid mass of masonry, or merely by one or two pillars of masonry and a filling-in of earth. It is rather difficult in a stone or brick bridge to say exactly what is the load on every portion of the arch, but it is guessed at, and a curve or link polygon, such as P h Q, Fig. 106 (inverted), drawn. It is shown in Art. 127 that in a stone or brick arch it is dangerous to have the arch so thin that the line P h Q (inverted) passes anywhere outside *the middle third* of the arch ring. Thus, in Fig. 107, we have a section of a stone arch, the various stones or *voussoirs*, as they are called, being separated by joints of mortar or cement. Now divide each joint into three equal parts and draw two polygons, $m\ m\ m$ and $n\ n\ n$, marking out the middle third of

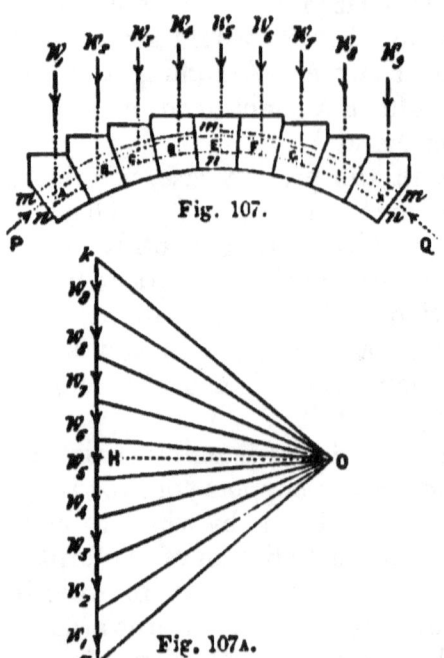

Fig. 107.

Fig. 107A.

every joint. Let us suppose we know the weight which each voussoir supports, including its own weight (it is usual to consider the arch as one foot deep at right angles to the paper), and let these weights be the weights w_1, w_2, &c., shown in Fig. 107. Now draw the force polygon, Fig. 107A; it happens to be all in one vertical line, the forces being all vertical. And now we come to the drawing of our link polygon, but we are stopped at the outset by not knowing w h a t i s t h e t h r u s t a t t h e c r o w n o f t h e a r c h. The pull at the middle of our suspension bridge chain was quite definite, but the thrust at the crown of the arch may be what we please, and the arch will remain stable if the link polygon which we draw never passes outside the middle third of any of the joints.* Suppose we draw any symmetrical link polygon to begin with, by bisecting $a\,k$ in H (Fig. 107A), draw H O horizontal, and take O anywhere we please. Now O H will be the thrust in the crown of our arch, if this link polygon is the correct one. Join O a, O 1 2, O 2 3, &c. Now start from any convenient point in w_5, Fig. 107, say E, within the space which contains the middle thirds of all the joints. Draw E D, Fig. 107, parallel to O 4 5, Fig. 107A; draw D C, C B, B A, A P, in succession parallel to the corresponding lines in Fig. 107A, and so also for E F, &c., to K Q. If any of the lines so drawn passes outside the space $m\,m$, $n\,n$, you must choose some other point E to begin at, and if you find that no choice of E will allow the link polygon to lie altogether within the space $m\,m$, $n\,n$, then you must choose another pole, O, in Fig. 107A, until at length you find, as in the figure, a link polygon, P E Q, which cuts within the middle third of every joint. The lengths of the lines in Fig. 107A tell us the forces acting at the joints of Fig. 107. Thus O a, Fig. 107A, is the force P A, Fig. 107, the resistance of the abutment of the bridge.

* It is obvious also that the link polygon wherever it crosses a joint must make an angle so near a right angle with the joint that there can be no slipping or rupture by shearing there.

Again, the length of o 4 5 is the force acting in the direction E D between the stones E and D.

157. Professor Fuller has made the work of drawing such a link polygon very easy. It can be shown that if a number of link polygons are drawn in Fig. 107 for different lengths, o H, Fig. 107A, then the vertical distances between the points A, B, C, D, &c., are in the same proportion in all the link polygons.* Beginning with the first load w_1, draw (Fig. 108) A 1' 2' 3' 4' 5' E, the half of any link polygon corresponding to P A B C D E in Fig. 107. Divide A S into any number of

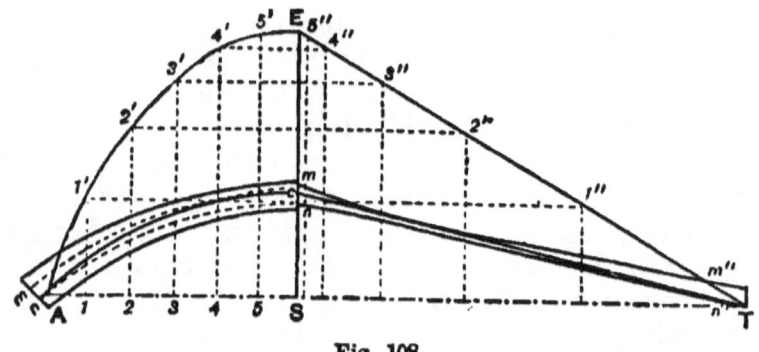

Fig. 108.

equal parts; I choose six. Erect perpendiculars at 1, 2, 3, 4, 5, s. Draw horizontal lines from 1', 2', 3', &c.; draw any inclined straight line of convenient length, E T; draw vertical lines from 1", 2", 3", &c. From the points where the verticals 1 1', 2 2', &c., cut $m\ m$ and $n\ n$, draw horizontals to cut the corresponding verticals from 1", 2", 3", &c. Join the points so found by the curves $m\ m''$ and $n\ n''$, then, just as the straight line E T represents a link polygon, $m\ m''n\ n''$ represents the area bounding the middle third of all the joints, and any link polygon will be represented on the right hand side by a straight line.

* See Art. 143. Each link polygon is really a diagram of bending moment supposing the loads were acting on a horizontal beam, and the scale of each diagram is proportional to the scale of O H.

Now draw a straight line lying altogether within the space $m\ m''\ n''\ n$. If you can draw several, then *draw that one which is steepest*, in this case c T. Project this over to the left hand side, and you will find that you have the link polygon, which supposes *the least thrust at the keystone*. The corresponding force polygon has its o н less than the o н of A E in the proportion s c to s E. The proof of this is easy. When the arch is one of wrought or cast iron, we have to find in just the same way the link polygon which falls nearest the axis of the rib everywhere. But we are not now bound to the middle third of the section, because the iron will withstand tension as well as compression. Art. 127 indicates how, when you have found the link polygon which passes most nearly through the middle of the iron rib everywhere, you can calculate the strength at every section.

158. In any arch which abuts against its supports at a plane surface, as a masonry arch does, the supporting forces and the link polygon are really indeterminate, although masonry arches in their settlement generally allow us to assume that the rule given in Art. 157 is true. In **iron arches** it is very important to have more definite information, and this is afforded by hinging the ends of the arch to the abutment. At a frictionless hinge the supporting force must pass through the centre of the hinge, and in such a case, if we assume that there is absolutely no yielding in the abutments, the bending in the arch everywhere will be such that the horizontal motions everywhere get equalised. This gives a simple rule to find the correct link polygon or line of resistance for an iron arch, for which the reader must be referred to Professor Fuller's paper on the "Curve of Equilibrium for a Rigid Arch under Vertical Forces."*

159. **Buttresses.**—To find the force which acts from one stone to another in a buttress, it is necessary to know the force acting on every stone from

* "Proceedings of the Institution of Civil Engineers," Vol. 40.

the outside, and also the weight of the stone. Find the resultant of these two forces for each stone, and draw the link polygon whose first side is the force on the top stone. In Fig. 109 F A B K P is the link polygon so drawn. Each side of it shows the resultant of the forces acting at every joint, and the length of the corresponding line in Fig. 110 shows its amount. Thus, the resultant of the forces acting at the joint S T is shown by the direction of the line B K, and its amount is shown by the length of the line $o\,t$ in Fig. 110.

If we see that any of the sides of the link polygon

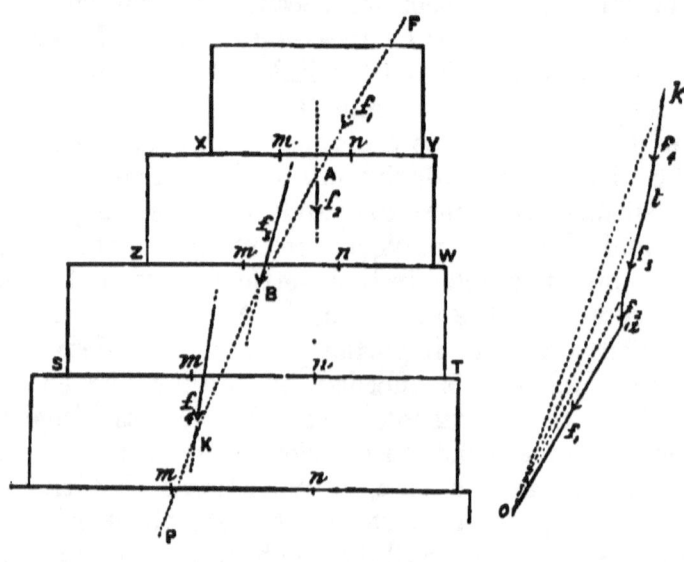

Fig. 109.   Fig. 110.

passes outside the middle third of the corresponding joint between two stones, we know that part of that joint will be subjected to tension, a condition to which we suppose that a common masonry joint ought not to be subjected.*

---

* In many cases it will be found well to magnify all the horizontal components of all the forces, magnifying the horizontal dimensions of all the stones in the same proportion. In this way the points in which each side of the link polygon cuts each joint may be found more accurately.

CHAPTER XVI.

SPIRAL SPRINGS.

160. As an example of the **bending** of a strip of material, which might have been considered after Art. 106, let us take the case of a **flat spiral spring**, such as the main or balance spring in watches. Let N P M (Fig. 111) be such a spring, fastened to a case at N, and to an arbor or axle at M. When no forces are acting on the spring it has a spiral shape. Suppose that in this case, at a point P, the radius of curvature is r_0, and that when the spring is partly wound up there is at P a radius of curvature r, then $\frac{1}{r} - \frac{1}{r_0}$ is the change of **curvature** at P, and we know that the bending moment which produced this change of curvature is equal to $EI\left(\frac{1}{r} - \frac{1}{r_0}\right)$, where E is the modulus of elasticity of the material and I is the moment of inertia of the cross section. (Thus, taking E at 36,000,000, if the breadth of the spring is 0·2 inch, and its thickness 0·03, then E I is 16·2.) Now suppose the arbor to have turned through the angle X O G (which I shall call A), from the unstrained condition. What are the forces acting on P M and the arbor? Whatever these forces may be, they must be in equilibrium. If these forces were changed there would be an alteration in the shape, but so long as these forces do not change, the shape and position of things do not alter. This is why we can apply to the spring, P M, and the arbor the **laws of forces acting on rigid bodies.** So long as P M O does not alter in shape, it obeys the laws of rigid bodies.

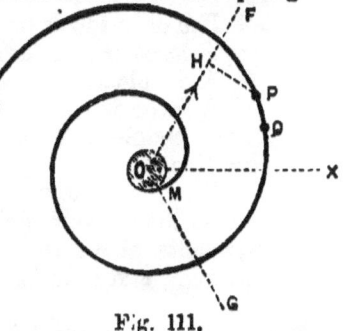

Fig. 111.

161. Now, the **forces acting on the arbor** may be very numerous—pressure of the pivots, pull of the fuzee chain, or pressure of teeth of wheels—but whatever they may be, we know that they can be represented by one force acting at o, the centre, together with a couple, c. If the spring is not in contact with the top or bottom of its case, and if the coils are not in contact with one another, no other forces act on the spring, M P, except at P. The particles of steel on one side of the section at P are acting on the particles on the other side; but whatever the forces at each of the particles may be, we know that the total effect at P is the same as that of one force and one couple. We cannot easily say what the force is, but if r is the radius of curvature at P, and if r_0 was the radius of curvature at P when the spring was unstrained, then the couple at P is what we have already called the bending moment.

$$\frac{Ebt^3}{12}\left(\frac{1}{r} - \frac{1}{r_0}\right)$$

Let us suppose, for simplicity, that the spring is everywhere of the same breadth and thickness, and let us use the letter e instead of

$$\frac{Ebt^3}{12}$$

which is now, of course, the same everywhere. The couple at P is then

$$e\left(\frac{1}{r} - \frac{1}{r_0}\right)$$

The only forces acting on P M O are—
A force at o, of amount f, in the direction o P, say.
A couple at o whose moment is —c.
A force at P.
A couple at P whose moment is given above, and is positive.

Now, we know that the sum of the moments of all the forces about any point must be nothing. Take all the moments about the point P. The force at P has then no moment, and is to be neglected, and we have

$$-f \times \text{P H} - c + e\left(\frac{1}{r} - \frac{1}{r_0}\right) = 0$$

In fact,

$$e\left(\frac{1}{r} - \frac{1}{r_0}\right) = c + f \cdot \text{P H}.$$

Let P Q be a short distance measured from P along the spring. Multiply every term of the above equation by P Q, and we find

$$e\left(\frac{\text{P Q}}{r} - \frac{\text{P Q}}{r_0}\right) = c \cdot \text{P Q} + f \cdot \text{P H} \cdot \text{P Q}.$$

FLAT SPIRAL SPRING.

Now, when P Q is very small it may be regarded as the arc of a circle whose radius is r; consequently $\frac{PQ}{r}$ simply means the angle between the radius, or normal at P, and the normal at Q; in fact, it means the small angle which the tangent at P makes with the tangent at Q.

Thus

$$\frac{PQ}{r} - \frac{PQ}{r_0}$$

simply means the change which has occurred in the angle, between the direction of the spring at P and the direction at Q. If now instead of considering what occurs at the point P, we take the point Q, we shall get just a similar equation for another little length of the spring. Suppose we do this for every short length of the spring, and add up our results; we shall find that the sum of all terms such as

$$\frac{PQ}{r} - \frac{PQ}{r_0},$$

means the change which has been produced in the angle, between the tangents to the spring, at its two ends. Thus, suppose the arbor has turned through the angle A, and suppose that, whether or not the point of fastening at N has been moved, the direction of the spring at N has on the whole changed through an angle B; then we find that the sum of all the above-mentioned terms amounts to A − B. (A may be called the amount of winding up of the spring; B may be called the amount of yielding in the fastening to the case.) Hence the sum of all the left-hand sides of all such equations as the above is e (A − B).

Now let us consider the right-hand sides of the equations. Evidently the sum of all such terms as C × P Q will be C × length of spring; say C l. The sum of all such terms as f × P H × P Q is, as you will find in any elementary book on mechanics, equal to f multiplied by the length of the spring multiplied by the perpendicular distance of the centre of gravity of the spring from the line O F. This is, of course, the length of the spring multiplied by the moment of the force f about the centre of gravity of the spring. Summing up our results, we find that if the force on the arbor through the pivots, &c., has a moment about the centre of gravity of the spring of the amount G, if the length of the spring is l, if the angle turned through by the arbor from the unstrained position is A, and if B is the angular yielding at N, and c is the couple with which the arbor tends to unwind itself, then

$$\theta(A-B) = cl - Gl;$$

or

$$c = \frac{\theta}{l}(A-B) + G.$$

The term G depends on the position of the centre of gravity of the spring.

162. If the coils are numerous, each will be nearly circular, and the centre of gravity of the spring will nearly be at O, and G becomes insignificant, so that the equation becomes

$$c = \frac{\theta}{l}(A-B).$$

If the spring is so rigidly fastened at its ends that there is no change of direction relatively to the barrel

$$c = \frac{\theta}{l}A,$$

and the couple exerted by the spring, in trying to unwind itself, is simply proportional to the amount of turning of the arbor, or the amount of winding up. If, then, the centre of gravity of the spring always remained in the centre of the arbor, and if the spring were rigidly fastened at N and M, we should have the couple exerted simply proportional to the angle of winding; and this is the condition for perfect isochronism in the balance spring. I need hardly say that this condition can never be perfectly satisfied. If we use a fuzee, the main spring may be fastened as we please; but suppose we want the couple exerted by the spring to be nearly constant, for various amounts of winding up, it is evident that the angle B ought to increase as fast as A; that is, there ought to be a very considerable amount of yielding in the fastening of the spring to its case. The same effect will be produced by exerting considerable pressure on the arbor at its pivots, or in some way causing the arbor and its case to be not quite concentric with one another.

The watchmaker's usual plan to get moderately good isochronism is to make one of the above errors tend to correct another; that is, by allowing a greater yielding or greater stiffness of the outer attachment counteract the results due to centre of gravity of the spring not remaining exactly in the axis of the balance.

163. Thus we see that by applying the law given in Art. 161 to the case of a flat spiral spring fastened to a case at its outer end, N, and to an arbor or axle at its inner end, M, we find that if the spring is riveted firmly both at N and M, and if it is so long and its coils so nearly circular that its centre of gravity is always nearly in the centre of the axle, then, when partly wound up, the spring tends to unwind itself with a **turning moment** which is **proportional to the amount of winding up.** This is the case in the balance spring, and it is this condition that gives to the balance its character of taking almost exactly the same time to make a small swing as to make a great one. (See Art. 180.) When the end N is not riveted, but merely hinged or fastened in any way that will allow it to turn about N, the unwinding tendency is not proportional to the amount of winding up; it is proportional to the difference between the angle of winding and this angular yielding at N. If the strip is everywhere of the same breadth and thickness, the unwinding tendency is proportional to the moment of inertia of its own section—that is, to its breadth and to the cube of its thickness; it is also proportional to the modulus of elasticity of the material used, and is inversely proportional to the total length of the strip. Suppose you wind a cord round the barrel or case containing a mainspring of a watch whose arbor is fixed firmly, and, using a scale-pan with weights, you find the turning moment of the spring for various amounts of winding up. If you plot your results on squared paper, you will find that the points lie in a curve like A o, B o, C o, or D o of Fig. 112, whereas for a balance spring we should get nearly a straight line through o.

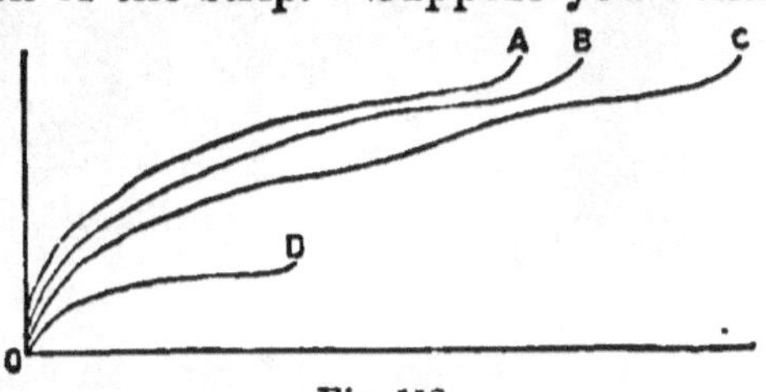

Fig. 112.

In Fig. 113 is represented an instrument which I have

been in the habit of using in my laboratory, to show the connection between the turning moment and the angular winding in a flat spiral spring. Different weights used at the end of the string give different readings of the

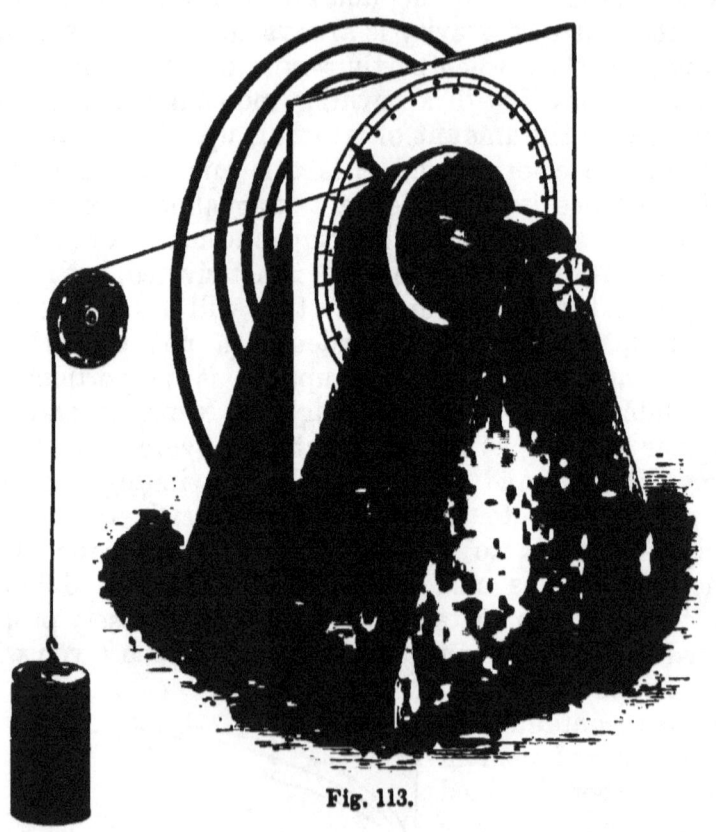

Fig. 113.

pointer. By means of such an apparatus, we are enabled to verify the laws described above. When we have performed one set of experiments with a spring, another set may be made on the same spring with its length diminished or increased by means of the arrangement for clamping, shown in Fig. 130. In this way we can experiment with springs of different breadths and thicknesses, as well as of different materials.

CYLINDRIC SPIRAL SPRING.

164. The flat spiral spring just considered is a case of the bending of a strip of steel along its entire length. I will now take up a case in which there is no bending. Fig. 114 shows a **cylindric spiral spring*** whose coils are very flat. Besides its own weight, it is acted upon by two equal and opposite forces in the direction of its axis, the supporting force at N and a weight at M. Now let us consider the equilibrium of the portion of the spring from any point P to M. Suppose the wire cut at P by a plane passing through the axis; this section will be more and more nearly a cross section normal to the axis of the wire, as the spirals are more and more nearly horizontal. Let us regard it as a normal cross section of the wire. Now, whatever may be the stresses at this cross section, they must balance all the other forces acting on P M—namely, the force F at M, which is axial, and the weight of P M, which is very nearly axial. If we neglect the weight of P M, we have only to balance the force F acting at M. To do so we evidently need a shearing force, F, at P, distributed over the section, and a twisting torque which is equal to F . P H. It is easy to show that the shear is of much less importance than the torsion. Again, since P H is the same for every part of the spring, every section of the wire is acted on by the same twisting couple, just as the shaft of Fig. 46, or the wire of Fig. 43, and its strength is calculated in the same way. Now, what is the amount of motion at M in consequence of this twist? As the wire is everywhere twisted, just as if it were a straight wire fastened at one end whilst at the other end there were a force, F, acting at the

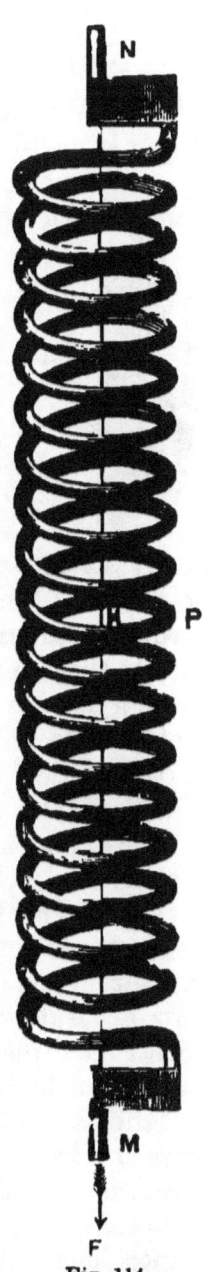

Fig. 114.

---

* Our authority on this subject is a paper by Professor James Thomson in the Cambridge and Dublin Mathematical Journal, Nov., 1848.

end of an arm whose length is equal to P H the radius of the coils of the spring, the amount of the motion of M is just the same as the motion of the end of such an arm attached to the straight wire.

165. We have, then, the following pretty illustration (Fig. 115), which serves to keep the rule for spiral springs in our memory. Let two pieces of the same wire of the same length be taken; one of them kept straight, fixed

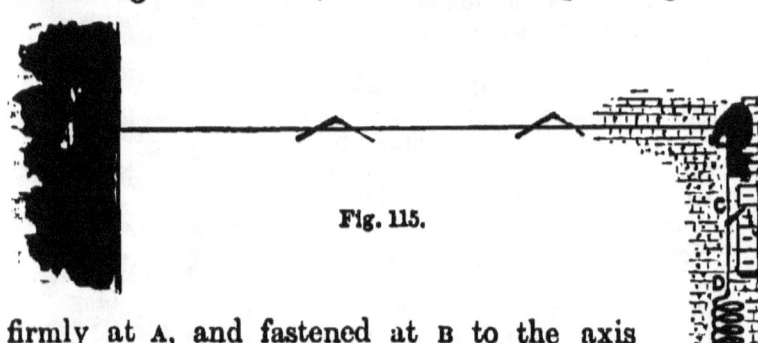

Fig. 115.

firmly at A, and fastened at B to the axis of a pulley which can move in bearings. A cord, C, fastened to the rim of this pulley, carries the upper end of a spiral spring, D E, formed of the other piece of wire, the diameter of its coils being equal to the diameter of the pulley. Evidently, if a weight, W, is placed in the scale-pan, a point E gets just double the motion of a point C, for E gets C's motion as well as the lengthening of the spring. The scales F and G and the little pointers are for the purpose of making exact measurements. It is interesting to note how accurately the law is fulfilled, even in a roughly-constructed piece of apparatus such as any one may easily put up for himself.

166. *Exercise.*—A spiral spring of charcoal iron spring wire, 0·1 inch diameter, 21·6 inches long, its coils having a radius of 1·3 inch, is extended by a weight of 10 lbs. Supposing that a piece of wire of the same material 1 inch long and 0·05 inch diameter, gets a twist of 24·2 degrees with a twisting moment of 2 inch-pounds, what is the extension of the spring? We see that if the trial wire were of twice the diameter the twist would be 24·2 ÷ 16, or 1·51 degree, and with a twisting moment of 13 inch-pounds, which is 6·5 times as great, the angle would be 9·82 degrees, and on a wire 21·6 times as long would be 212· degrees, or 3·7 radians, and the arc of a circle whose radius is 1·3 inch, subtending this angle is 3·7×1·3 or 4·8 inches, the answer.

167. In designing a cylindric spiral spring it is very important to know the **greatest elongation** it will bear without taking a permanent set. If the material has internal strains given to it during its manufacture—and this it is very difficult to prevent in steel springs, unless great care is taken in tempering, and it is almost impossible to prevent in brass springs, because the elasticity added in manufacture is often regarded as a necessary quality which ought not to be destroyed by any annealing process—in this case the reader must keep in mind the considerations of Art. 96. Otherwise, let s be the greatest shearing stress per square inch which the material can resist without getting a permanent set. Let m be the greatest twisting moment which a round wire of radius r can bear without getting a permanent set, we see from Art. 93 that

$$m = \tfrac{1}{2} \pi s r^3. \quad \text{...............(1)}$$

Now s will be approximately known from Table III, or m may be found by experiment for a given wire by any person who wishes to make a spring; and whether m or s is used in a formula, you now know how to calculate one when given the other and the size of the wire. If, then, we have a spring made of wire whose radius is r, and if the radius of the coils as measured to the centre of the wire from the axis of the spring is a, we see that when W is the greatest weight with which the spring may be elongated without producing a permanent set,

$$W = \frac{m}{a} = \frac{\pi s r^3}{2a} \quad \text{..................(2)}$$

being independent of the length of wire employed.

From Art. 92 we see that if N is the modulus of rigidity of the material, A the greatest angular twist in radians which we can give to a wire of radius r inches and length one inch, and m the twisting moment which produces this twist, then

$$A = \frac{2m}{\pi N r^4};\quad\quad\quad\quad\quad\quad\quad\quad (3)$$

m being what we have previously measured or calculated. N is approximately known for a material from Table III., or A may be found by experiment for a given wire; and whether A or N is used in a formula, you now know how to calculate one when given the other.

Putting the result of our reasoning in Art. 164 into an algebraic form, we see that a load, w, will elongate the spring by the amount

$$\frac{2 w l a^2}{\pi N r^4},$$

and hence the greatest elongation which can be given to the spring without its getting a permanent set is

$$l A a, \text{ or } \frac{2 l a m}{\pi N r^4}, \text{ or } \frac{l a s}{N r}\quad\quad\quad\quad (4)$$

Combining (2) and (4), we see that when a spring is stretched to its elastic limit, the mechanical energy stored up in it, which is called its "*resilience*," being half the product of w into the elongation, is

$$\tfrac{1}{2} m l A, \text{ or } \frac{\pi s^2 r^2 l}{4 N}, \text{ or } \frac{l m^2}{\pi N r^4}\quad\quad\quad (5)$$

168. Many interesting methods may be taken to express in words the meanings of these results. Thus, the second expression in (5) shows that the **work which we can store up** in a spiral spring is simply proportional to the weight or quantity of material in it. It would be easy to show that we can store more energy in a spring formed of wire of circular section than in one of equal weight of the same material whose wire has any other than a circular section.

169. The following **readings** of **our formulæ** may prove to be useful:—1st. If r, the radius of the wire, and a, that of the coils, be fixed, the elongation produced by any weight, w, will be proportional to l, the length coiled up to form the spring. 2nd. If a wire of a certain length and radius be given to form a spring, the elongation produced by a certain weight, w, will be proportional to the square of the radius which we may adopt for the coil. 3rd. If the radius of the wire be fixed, and the length of the spring when closed, so that the coils may touch one another, or, what is the same, the number of coils be also fixed, l must be proportional to a, and therefore the elongation due to a weight, w, will be proportional to the third power of the radius which we

may adopt for the coil. 4th. If the length of the wire and the radius of the coil be fixed, the elongation due to a weight, w, will be inversely proportional to the fourth power of the radius of the wire which we may adopt. 5th. With a given weight of metal and a given radius of the coil, the elongation due to a weight, w, will be proportional to l^3, or inversely to r^6, since l must be proportional to $\frac{1}{r^2}$.

We see that the ultimate elongation is—1st, proportional to the length of the wire, if the radius of the wire and that of the coil be fixed. 2nd, proportional to the radius of the coil, if the length and the radius of the wire be fixed. 3rd, Inversely proportional to the radius of the wire, if the length of the wire and the radius of the coil be fixed.

It will be found that a weight hung at M (Fig. 114) will tend to turn as the spring lengthens, unless the coils of the spring are very flat. This is due to the fact that the cross sections of the wire are really subjected to a little bending as well as torsion.

170. We can cause the strain in such a spring to consist altogether of bending, if, without exerting any axial force such as I have shown in Fig. 114, we exert a couple about the axis such as we exerted on the wire in Fig. 43. The wire in Fig. 43 would be twisted, but the wire in Fig. 114 is subjected everywhere to bending without any twisting or with only a very little twisting, due to the fact that the coils are not perfectly flat.

If a_0 is the radius of the coils to the centre line of the wire when unstrained, and the length of the coiled wire is l, then the number of coils multiplied by the circumference of each is the total length, so that the number of coils is $l \div 2\pi a_0$. If now the moment of inertia of the cross section of the wire about the axis through its centre about which it bends is I, and if M is the moment which acts at the unfixed end of the spring to twist it, then the new radius, a, of every coil is obtained from our knowledge of the fact given in Art. 160.

$$\text{M} = \text{E I} \times \text{change of curvature},$$

or $\text{M} = \text{E I} \left(\frac{1}{a} - \frac{1}{a_0} \right)$,

E being the modulus of elasticity of the material.*

---

* I is $b\,d^3 \div 12$ for a wire of rectangular section, d being the dimension in inches of the section, measured radially out from the axis of such a spring; I is $\pi d^4 \div 64$ for a wire of circular section of diameter d.

M

From this we have

$$\frac{M}{EI} = \frac{1}{a} - \frac{1}{a_0}, \text{ or } \frac{Ml}{EI} = \frac{l}{a} - \frac{l}{a_0}, \text{ or } \frac{Ml}{EI} = n - n_0,$$

where n_0 was the original number of windings, and n is the new number of windings. But one additional winding means that the unfixed part of the spring has moved through 360 degrees, or 2π radians; and hence if A is the angular motion of the unfixed end of the spring in radians,

$$A = \frac{2\pi Ml}{EI}.$$

We see that it does not depend on the radius of the coils. For a spring of round wire of diameter d, the angular motion due to a turning moment, M, is

$$A = 128\, Ml \div E d^4.$$

171. From these considerations it is evident that a **spiral spring** like Fig. 115 when it lengthens under the action of a weight, has all its wire subjected to torsion. The spring itself is extended, but the wire of the spring is twisted. Again, if we subject the spring to torsion as a whole, the strain really going on in the wire is a bending strain. Usually, a spiral spring, as its coils are not perfectly flat, has its wire subjected to torsion principally, and a little bending as well, when the spring is extended; and when the spring is twisted as a whole its wire is mainly subjected to bending, but there is also a little twist in it. **The extension of a spiral spring** is proportional to the pulling force, and also to the length of the wire and to the diameter of the coils; it is inversely proportional to the fourth power of the diameter of the wire if the wire is round. **The twist given to a spiral spring** is proportional to the moment of the twisting forces, it does not depend on the size of the coils; it is proportional to the length of wire, and inversely proportional to the fourth power of the diameter of the wire if the wire is round.

CHAPTER XVII.

PERIODIC MOTION.

172. When, after a certain interval of time, a body is found to have returned to an old position, and to be there moving in exactly the same way as it did before, the motion is said to be **periodic**, and the interval of time that has elapsed is said to be the **periodic time** of the motion. Thus, if a body moves uniformly round in a circle, the time which it takes to make one complete revolution is called its periodic time.

173. When a body moves uniformly in a circle, as, for instance, the bob of a *conical pendulum* (see GLOSSARY, Art. 236), if we look at it from a point in the plane of its circle, it seems merely to swing backwards and forwards in a straight line. Thus, it is known that Jupiter's satellites go round the planet in paths which are nearly circular, but a person on our earth sees them move backwards and forwards almost in straight lines. Now if we were a very great distance away from the bob of a conical pendulum in the plane of its motion, we would imagine it to b moving in a straight line, and the motion which it would appear to have—slow at the ends of its path, quick in the middle—would be a **pure harmonic motion**. To get an exact idea of the nature of this motion—in fact, to define what I mean by pure harmonic motion—draw a circle A O' L O" (Fig. 116), and divide its circumference into any even number of equal parts. Draw the perpendiculars B' B, C' C, &c., to any diameter. Now, if we suppose a body to go backwards and forwards along A O L, and if it takes just the same time to go from A to B as from B to C, or from any point to the next, then

its motion is said to be a pure harmonic motion. This sort of motion is nearly what we observe in Jupiter's satellites; it is almost exactly the motion of the bob of any long pendulum or the cross head of a steam-engine; it is the motion of a point in a tuning-fork, or a stretched fiddle-string when it is plucked aside and set free; of the weight hung from a spring balance when it is vibrating; of a cork floating on the waves in water; and of the free end of a rod of metal when the other end is fixed in a vice and the rod is set in vibration; it tells us in all these cases the nature of the motion, when such motion is of its simplest kind. Thus, for example, a cork floating on water may really have a very complicated motion, but if the wave in the water is of its simplest kind, the cork goes up and down with a pure harmonic motion. If you study the figure which you have drawn, and then watch the vibration of a very long pendulum, you will learn about this kind of motion what cannot be learnt by reading.

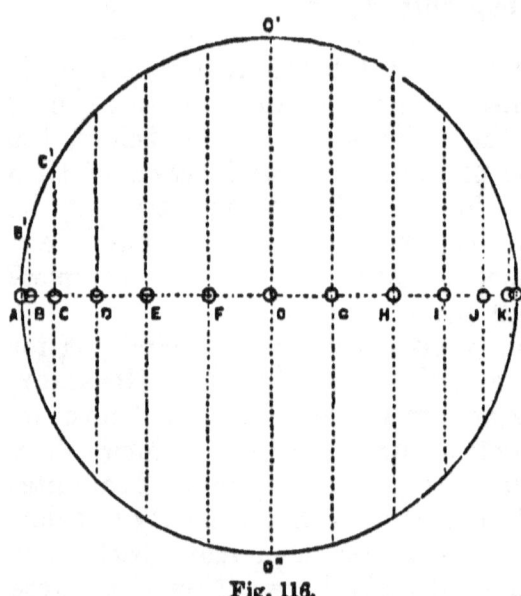

Fig. 116.

174. Now let me suppose that the body takes one second to go from A to B, or from B to C, or from any point to the next in Fig. 116. Then the *length of* A B *in inches represents the* average velocity between the points A and B, and in the same way we get the average

velocity anywhere else. Thus, in the figure from which the woodcut is drawn I find

Velocity from	A to B	B to C	C to D	D to E	E to F	F to O	O to G	G to H	H to I	I to J	J to K	K to L
Is in inches per second	0·34	1·00	1·59	2·07	2·41	2·59	2·59	2·41	2·07	1·59	1·00	0·34

175. You will observe that the velocity increases as the body approaches the middle of the path, and diminishes again as it goes away from the middle. Now the increase in the velocity of a body every second is called its acceleration, and I want you to observe **what is the acceleration at every place**. You see that the velocity changes from ·34 to 1·00 near B in one second—that is, the acceleration near B is ·66 inch per second per second. Similarly subtracting 1·00 from 1·59 we find the acceleration at C to be 0·59, and so on. Now make a table of these values, and place opposite them the distances of the points B, C, &c., from the centre. In this way I find from my figure the following Table of Values :—

Distance from o to	Acceleration at	Displacement divided by Acceleration.
B is 9·66	B is 0·66	14·6
C is 8·66	C is 0·59	14·7
D is 7·07	D is 0·48	14·7
E is 5·00	E is 0·34	14·4
F is 2·59	F is 0·18	14·4
O is 0	O is 0	—
G is 2·59	G is 0·18	14·4
H is 5·00	H is 0·34	14·4
I is 7·07	I is 0·48	14·7
J is 8·66	J is 0·59	14·7
K is 9·66	K is 0·66	14·6

From this it is evident that when the distance of a point from the centre is divided by the acceleration at the point, you get about 14·6 in every case-that is, the acceleration at a place is proportional to the distance from the centre. This curious property is characteristic of the kind of motion which I am describing. If, again, you draw a number of figures, such as Fig. 116, and divide the circles into very different numbers of equal parts, you will find that in every case the following law is true:— The periodic time of a pure harmonic motion—that is, the time which elapses from the moment when the body is in a certain condition until it gets into exactly the same condition again—is equal to 6·2832 multiplied by the square root of the ratio of displacement to acceleration given in the third column of the above Table. Thus, in the Table we find the mean value of the ratio (adding all the quotients and dividing by their number we get 14·56) to be, let us say, 14·6. Now the square root of 14·6 is 3·82, and this multiplied by 6·2832 is 24 seconds, which we see by inspection is the periodic time in Fig. 116.

The acceleration is always towards the middle point—that is, whilst a body is leaving the middle, its velocity is being lessened, when it is approaching the middle its velocity is being increased. The velocity at the middle is equal to the uniform velocity in the circle from which we imagine the harmonic motion to be derived—that is, the velocity in the middle is equal to 3·1416 times the distance A L divided by the periodic time.

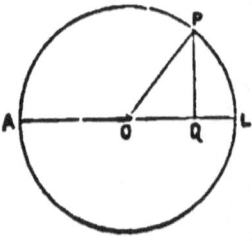

Fig. 117.

Suppose the body to be at Q, Fig. 117, moving with a pure harmonic motion in the path A O L. Describe the circle, draw Q P perpendicular to A L, then P is the position of a body which has corresponding uniform circular motion. The acceleration at Q is equal to the resolved part along O L of the centripetal acceleration at P in the direction P O, which is known to be $v^2 \div$ P O where V is the uniform velocity of P.

The resolved part of this in the direction Q o is evidently obtained by multiplying by Q o and dividing by P o, and we get for the acceleration of Q, $v^2 \times Q o \div P o^2$, so we see that the acceleration of Q is proportional to Q o. The acceleration at L or A, the ends of the path, is of course greater than anywhere else, being $v^2 \div P o$.

If at any place, Q, we divide the displacement by the acceleration, we get $Q o \div \frac{v^2 \cdot Q o}{P o^2}$, or $P o^2 \div v^2$, and as v is the circumference of the circle $2 \pi \cdot P o$, divided by the periodic time T, we have

$$\left. \begin{array}{l} \text{Periodic time of a pure} \\ \text{harmonic motion} \end{array} \right\} = 2 \times 3\cdot 1416 \sqrt{\frac{Displacement}{Acceleration}}.$$

We see then that if the force acting on a body and causing it to move is always proportional to the distance of the body from a certain point, and acts towards that point, the body gets a pure harmonic motion, and we have a **rule for finding the periodic time.**

176. *Example.*—In Fig. 118, A is a ball of lead weighing 20 lbs. carried by means of a spiral spring whose own weight may be neglected, let us suppose. Find by experiment how much the spring lengthens when we add 1 lb. to the weight of A or shortens when we subtract 1 lb. from the weight of A. Let it lengthen or shorten 0·01 foot. Evidently, if ever A is 0·01 foot upwards or downwards from its position of rest, it is being acted upon by a force of 1 lb. tending to bring it to its position of rest. We know also (see Art. 171) that if A is 0·02 foot or 0·03 foot above or below its place of rest, there is a force of 2 or 3 lbs. trying to bring it back. We see then that the up and down motion of A must be pure harmonic. When the displacement is, say 0·02 foot, the force acting on A is 2 lbs., and the acceleration of A is force $2 \div$ mass of A,

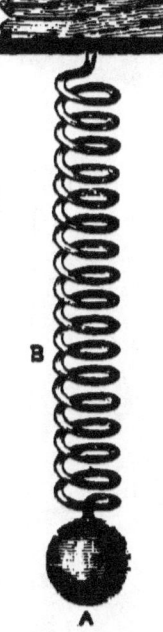

Fig. 118.

and as the mass of A is 20 ÷ 32·2, or 0·621, the acceleration of A is 3·22 feet per second per second when it is displaced 0·02 foot from its middle position. Now employing the rule given above, divide 0·02 by 3·22 and extract the square root, then multiply by 6·2832, and we get 0·495 second, or about half a second as the **periodic time of the swinging ball.** If you make experiments you will find that, unless the coils of the spring are very flat indeed, and the rigid support of A exactly in the axis, the ball has a tendency to turn and to vibrate laterally, which will somewhat disturb your observations if you are making careful measurements of the length of swing.

177. *Example*.—The Simple Pendulum.—A simple pendulum consists in an exceedingly small but heavy body suspended by means of a long thread whose weight may be neglected, capable of swinging backwards and forwards in short arcs. Thus, in Fig. 119, S is the point of suspension, S P a silk thread, P a small ball of lead. P will move backward and forward along the path A O L with a motion which is pure harmonic, provided the thread is so long and A L so short that A L may be regarded as nearly straight, because the force acting on the ball at any time in the direction of its motion is proportional to the distance of the ball from O. To show that this is so, resolve the vertically acting weight of the ball in the direction of its motion along A O. You find that it is not quite proportional to A O unless A O is very nearly straight, but if this slight discrepancy is neglected the force urging the ball towards O is the weight of the ball multiplied by O A and divided by S A, the distance from

Fig. 119.

the point of support to the centre of gravity of the ball. As a matter of fact, the nature of the vibration does not depend on the weight of the ball; but, to fix our ideas, let us suppose that the weight is 2 lbs., then the mass of the ball is $2 \div 32\cdot 2$, and acceleration along A O is the force $\div$ mass, or $\frac{2 \times \text{A O}}{\text{S A}}$ the force $\div \frac{2}{32\cdot 2}$ the mass, so that the acceleration is $32\cdot 2 \times \text{A O} \div \text{S A}$.

Now our rule is to divide A O by the acceleration at A, and this gives $\frac{\text{S A}}{32\cdot 2}$; extract the square root, and multiply by 6·2832 for the periodic time of oscillation of the pendulum. The general rule for a simple pendulum swinging in short arcs is then:

$$\text{Time of a complete oscillation} = 6\cdot 2832 \sqrt{\frac{\text{length of pendulum}}{32\cdot 2}}$$

The time of *one swing* is half this. The number 32·2 expresses the effect of the force of gravity at London. At any other place on the earth's surface it would be different —that is, at different places on the earth a given pendulum has different times of oscillation. For instance, a pendulum taking 2 seconds for a complete oscillation at Paris, that is, taking 1 second for one swing, called a seconds pendulum, if swung at Spitzbergen would gain 94 seconds per day, and if swung in New York would lose 30 seconds per day, provided the pendulum did not alter in length in being taken from one place to the other. Evidently when a pendulum gets longer it oscillates more slowly; hence in summer, when the pendulum of a common house-clock expands with heat, it goes more slowly, and in winter it goes more quickly, unless the position of the bob is adjusted. A pendulum which is self-adjusting—that is, which is so constructed that it remains of the same length whatever be the temperature— is called a *compensation pendulum*.

178. *Example.*—In Fig. 120, B represents a strip of steel fixed firmly in a vice at C, with a heavy ball A

fastened at its free extremity. Find the force in pounds which will increase the deflection of A by 0·01 foot; say that it is 1 lb. We know that a force of 2 or 3 lbs. will

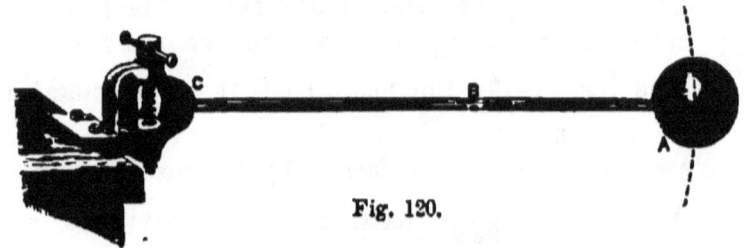

Fig. 120.

cause an increased deflection of twice or three times this amount, and as the force acting on the ball in any position is proportional to its distance from its position of rest, the ball will swing with a pure harmonic motion. If we can neglect the weight of the strip of steel, and if the ball is small in comparison with the length of the strip, its time of vibration may be calculated in exactly the same way as that of the ball in Fig. 119.

179. *Example.*—Suppose that B C (Fig. 121) is a **bent glass tube of uniform section** containing a liquid which can move without friction in the tube. If the liquid be disturbed so that the level is higher in B than in C, it will continue to swing about its position of equili-

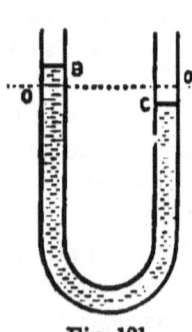

Fig. 121.

brium, that is, the position in which the liquid is at the same level in both limbs of the tube. Thus, if C is ·01 foot below the proper level, and B is ·01 foot above this level, the force which tends to cause the liquid to return to its proper level is twice the weight of the liquid O B. Suppose the weight of the liquid is 10 lbs. per foot in length of the tube, then the force acting on the liquid is ·02 × 10, or ·2 lb. If the whole length of tube filled with liquid is 6 feet, then the weight of liquid which has to be set in motion is 60 lbs., and its mass is $60 \div 32·2$, or 1·863; hence the acceleration is $·2 \div 1·863$, or 0·107 foot per second per second. The displacement is ·01, and, working by our old rule, displacement divided by acceleration is ·0935. The square root of this is ·3058, and multiplying by 6·2832 we get 1·92 second as the periodic time of the oscillation.

You will find it easy to prove that **the liquid swings in the same time as a simple pendulum whose length is half the total length of the liquid in the tube, and that it is the same whatever be the density of the liquid**—that is, whether it is mercury or water.

If w lbs. is the weight of liquid per foot in length of the tube, if d is the displacement o b or o c, the force causing motion is $2\,d\,w$. If a is the total length of liquid in the tube, the weight of liquid moved is $a\,w$, and its mass is $a\,w \div g$, if g is 32·2, which represents the effect of gravity.

Hence the acceleration is $2\,d\,w \div \frac{a\,w}{g}$, or $\frac{2\,d\,g}{a}$, and the displacement divided by acceleration is $d \div \frac{2\,d\,g}{a}$, or $\frac{a}{2\,g}$, so that the periodic time is $2\,\pi\,\sqrt{\frac{1}{2}\cdot\frac{a}{g}}$.

180. It will be observed that in all these cases of vibration of bodies there is a **continual conversion going on of one kind of energy into another.** At each end of a swing the body has no motion; all the energy is therefore potential, whether it is the potential energy of a lifted weight or the potential energy of strained material. In the middle of the swing the body is going at its greatest speed, and its energy is kinetic. At any intermediate place the energy is partly potential and partly kinetic, but the sum of the two remains always the same, excepting in so far as friction is wasting the total store. Now in time-keepers the **office of the mainspring** is to give just such supplies of energy to the balance as are necessary to replace the loss by friction; and we have to ask the question—At what part of the swing of a pendulum or balance can we give to it an impulse which shall increase its store of energy without disturbing its time of oscillation? The answer is this. If a blow is given to the bob of a pendulum when it is just at its lowest point, energy is given to the pendulum; we give it power to make a greater swing, but the time which it will take to make this greater swing is just the same as the time it would have taken for a smaller swing. This middle point is the only point at which we can give an impulse to the bob without altering the time of its swing. In the lever escapement, and in other detached escapements of watches, the impulse is always given just at the middle of the swing.

CHAPTER XVIII.

OTHER EXAMPLES OF PERIODIC MOTION.

181. When the periodic motion of a body is not pure harmonic, we find that by imagining the body to have two or more kinds of pure harmonic motion at the same time we can get the same result. Thus, it is known that a float, employed to measure the rise and fall of the tide by marking on a moving sheet of paper with a pencil, has a motion which is periodic and not pure harmonic. Thus, if horizontal distances represent the motion of the paper (unwound from a barrel by means of clockwork), and therefore represent time, and if vertical distances mean the rise or fall of water-level in feet, we get such a curve as is shown in Fig. 122.

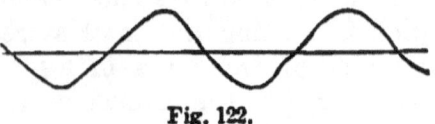

Fig. 122.

Now this is not a pure harmonic motion, for if you plot on squared paper the distances O A, O B, O C, O D, O E, O F, &c. (Fig. 116), for equal intervals of time, you will get a curve like Fig. 123, which is easily recognised, and is called a *curve of sines*. But it has been found that if you take certain curves of sines whose periodic times are—1, the semi-lunar day; 2, the semi-solar day, and some others, and draw them on squared paper, and add their ordinates together, you will get the curve shown in Fig. 122. In the very same way you can com-

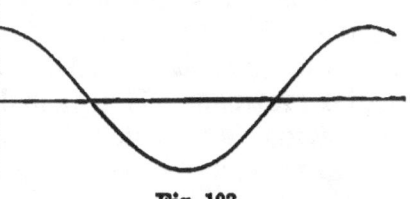

Fig. 123.

bine pure harmonic motions to arrive at any periodic motion. A good way of combining pure harmonic motions experimentally is to let a body hang from a string which passes over two or more movable, and the same number of fixed, pulleys. These pulleys are pivoted on crank pins, and their pivots are made to revolve at any desired relative speeds, and each gives to the body a pure harmonic motion by its action on the string. The body gets a motion compounded of the motions of the pulleys, and if it is an ink-bottle or pencil pressing on the paper on a revolving paper roller, we get a time curve of the periodic motion. This is the principle of the construction of Sir William Thomson's Tide Predicting Machine.

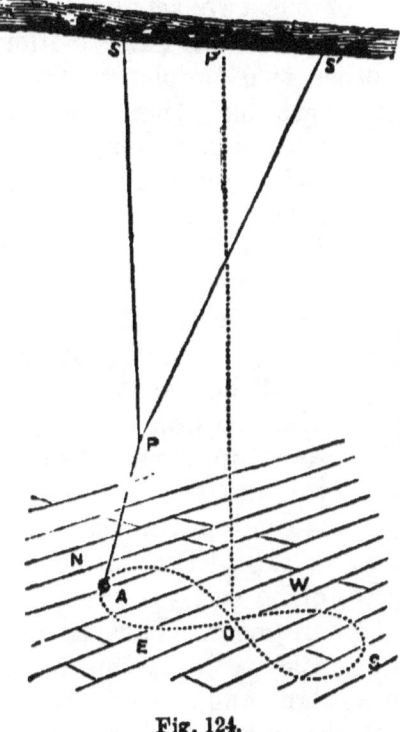

Fig. 124.

182. When a body can swing east and west under the influence of forces which have no tendency to move it except in a direction due east and west, and if forces acting due north and south can make it swing in their direction, then both sets of forces acting together on the body will give it a **motion compounded of the two simpler motions.** Thus, a ball A (Fig. 124) is suspended by a string, P A, which is knotted at P to two other strings, P S and P S', equal in length, and fastened at s and s'. The ball may swing in the direction E O W as if it were the bob of a pendulum hung directly from the ceiling at P', but it may also swing in the direction N O S at right angles to E O W, and if it does so it swings as if the point P were the fixed end of the pendulum A P. When it swings under the influence of the two sets of forces tending to make it move both ways at once, the motion of A is compounded of the other two simpler motions. If P A is one-quarter of the length O P', then the east and west swing takes twice as long as the north and south swing. If P A is one-ninth of O P', then the east and west swing takes three times as long as the north and south swing. The motion of A is sometimes very beautiful, and the experiment is easily arranged.

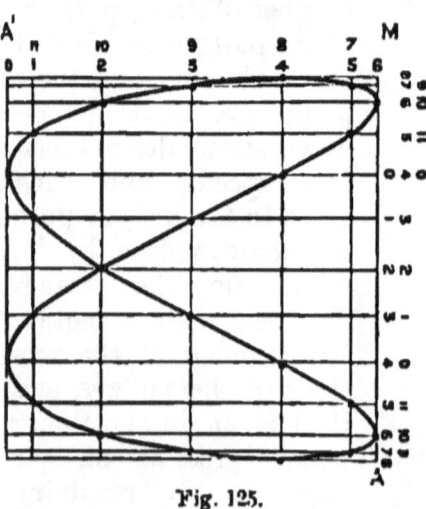

Fig. 125.

183. The motion is quite easily represented on paper. Thus, in Fig. 125, A' M is the north and south direction, and A M, at right angles to it, is the east and west direction. Let the points 0, 1, 2, &c., in each of these lines be found as in Fig. 116. Let the bob be supposed to go from 0 to 1 in A'M in the same time as

COMBINATION OF VIBRATIONS.

it goes from 0 to 1 in A M. You will observe that I have twice as many points in A M as in A′ M, showing a slower oscillation in the direction A M. You can begin to number

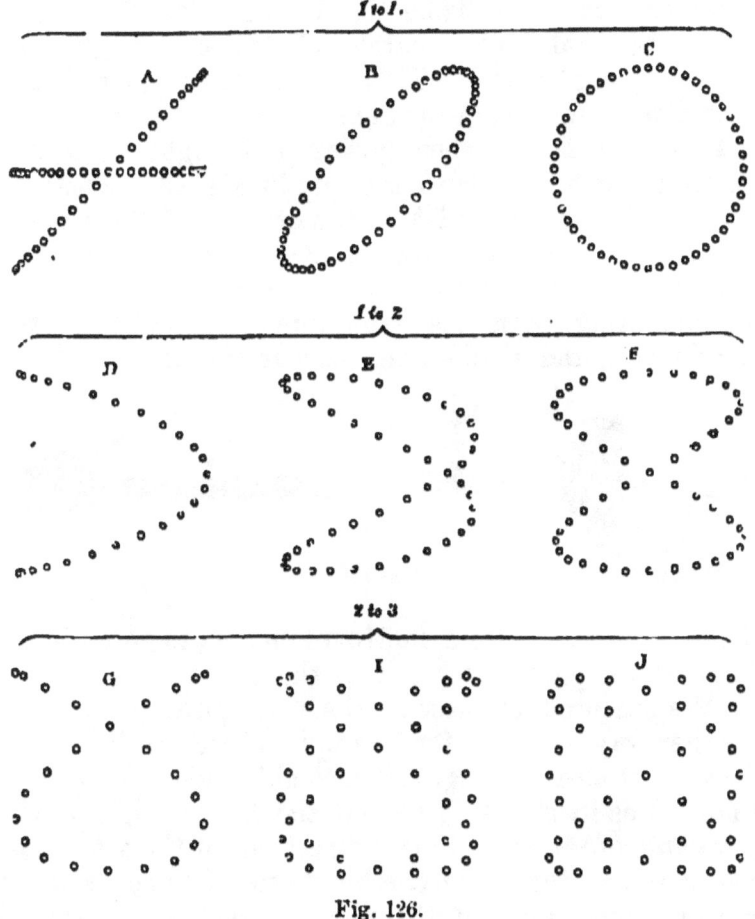

Fig. 126.

your points anywhere, remembering that when the bob completes its range it comes back again in the opposite direction. Now put marks where the east and west lines meet the north and south ones, drawn through corresponding points. It is evident that the curve drawn through these successive marks is the real path traced out by the ball when acted upon simultaneously by the two sets of forces urging it in a north and south, and an east and west direction.

If you have the same number of points in A'M as in A M, you will get a circle, ellipse, or straight line, as in A, B, C, Fig. 126. This represents the motion of a conical pendulum free to swing in every direction. Again, D, E, F, and many other curves that might be drawn, represent the case which I took up in Fig. 125, where one vibration is twice as quick as the other. If the time of vibration in A M is to the time of vibration in A' M as 2 to 3, we get curved paths like G, I, J, and so on. In experimenting with the pendulum, Fig. 124, it will usually be found that slight inaccuracies in the lengths

Fig. 127.

of the cords will cause a continual change to go on in the shape of the path traced out by the ball.

We can produce these motions by spiral springs, and in other ways. Thus, for example, if we use instead of the strip of steel, in Fig. 120, a combination of two strips, B and B', as in Fig. 127, so that the heavy bright bead A is capable of vibrating in two directions at the same time, you will get the same combinations of pure harmonic motions, depending on the point at which B is held in the vice C.

184. When a body has a periodic **rotational motion** about an axis like the balance of a watch or a rigid pendulum, we must no longer speak of the force causing motion, and the mass of the body, and the distance of displacement; but if we substitute for these terms, moments of forces, moment of inertia of the body and angle of displacement, we have exactly the same rule for finding the periodic time of oscillation. The periodic

time is 6·2832 times the square root of the angular displacement of the body at any instant, divided by the angular acceleration at that instant. And we know that angular acceleration may be calculated by dividing the turning moment acting on a body by the moment of inertia of the body. A point in the **balance of a watch** swings in circular arcs, but if you only take account of the distances which it passes through, and suppose it moved in a straight line instead of in the arc of a circle, the motion is very nearly pure harmonic. If there were no friction or other forces acting on the balance except the turning moment of the balance spring (see Arts. 162-3), and if the moment of the spring were always exactly proportional to the angular displacement of the balance, the motion would be pure harmonic.

We saw in Art. 161 that the turning moment of the spring is $\frac{E b t^3}{12 l} A$, if E is modulus of elasticity of the spring, b its breadth, t its thickness, and l its length, and if A is the angular displacement in radians.

Angular acceleration is this moment divided by moment of inertia I of the balance, or $\frac{E b t^3}{12 l} \frac{A}{I}$.

Hence, angular displacement A, divided by angular acceleration, is $\frac{12 l I}{E b t^3}$, so that the **periodic time of the balance** is

$$T = 6\cdot 2832 \sqrt{\frac{12 l I}{E b t^3}}.$$

Increasing the moment of inertia of the balance or the length of the spring makes the vibration slow. Increasing the breadth and, what is still more important, increasing the thickness of the spring makes the vibration quick. As we saw in Arts. 162-3 that our calculation of the turning moment of the spring is not quite right, that the dimensions of the balance and spring alter with temperature, and that above all the elasticity of the steel alters with temperature and with its own state of fatigue, the rule given in the note is not perfectly true, nor can any balance be regarded as taking exactly the same time for its oscillation in different lengths of arc. At the same time it is of great help to the watchmaker to know that with considerable, although not with perfect accuracy, the time of vibration of a balance is proportional to the square root of the length of the spring, and so on. For example, suppose the spring is 3 inches long, and the balance makes one swing in 0·251 second, now if he wishes it to make a swing in 0·25 second, he must shorten it in the ratio of ·251 × ·251 to ·25 × ·25, or in the ratio ·063001 to ·0625, so that the length of his spring ought to be 3 × ·0625 ÷ ·063001, or 2·976 inches— that is, it ought to be shortened ·024 inch. In the same way he can calculate the effect of adding little masses at any distances from the centre of the balance, so that its moment of inertia may be increased, and the balance made slower in its swing. The same law tells him how he can compensate the balance, so that when in summer the steel of the spring loses its elasticity, some of the mass of the balance will come nearer the centre, in order that the moment of inertia may diminish the same proportion.

185. **Compound Pendulum.**—The simple pendulum described in Art. 177 is not like the pendulums used in practice. In these the bob is not so small that we can consider it as a point; the long part is not a thread but a stiff rod of metal or wood, and there is usually a knife-edge for support, about which it can turn with little friction. In common clocks, however, the top end of the pendulum is a thin strip of steel held firmly in the chops, but the easy bending of this strip is such that we may imagine the pendulum to move freely about an axis. Employing our general rule of Art. 184 we find how to calculate the time of vibration. This compound pendulum vibrates in the same time as a certain simple pendulum, called the *equivalent simple pendulum*, whose length you ought to find by experiment.

Chap. XVIII.] COMPOUND PENDULUM. 195

In Fig. 128 let s be the *axis of suspension*, G the centre of gravity, and P a point in the continuation of the line S G such that S P is the length of the equivalent simple pendulum. Then P is called the *centre of oscillation*, and it is also known to be the *centre of percussion* of the pendulum (see Art. 200). It can be proved that if the pendulum be inverted and made to vibrate about a parallel axis through P, it will vibrate in exactly the same time as it does about S; and it was in this way, by inverting a pendulum which had two knife-edges, and adjusting these until the pendulum took the same time to vibrate about one as about the other, and then measuring the distance between them, that Captain Kater found the length of the simple pendulum which vibrates in a given time. This method is still employed in gravitation experiments everywhere to find the value of g which is 32·2 feet per second per second at London.

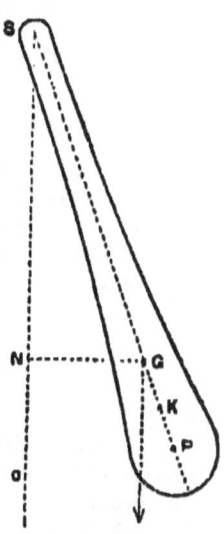

Fig. 128.

If S is the axis of suspension, G the centre of gravity, W the weight of the pendulum, then the moment with which gravity urges the pendulum to return to its position of rest is W × G N, but if the angle G S O be measured in radians, and if it is very small, this moment is almost exactly equal to W × S G × angle G S N. The angular acceleration is obtained by dividing this by I, the moment of inertia of the pendulum about S, and our rule becomes

$$T = 6·2832 \sqrt{\frac{\text{angle } G S N}{W \cdot S G \cdot \text{angle } G S N \div I}}$$

$$\text{or } T = 6·2832 \sqrt{\frac{I}{W \cdot S G}} \quad \ldots\ldots\ldots\ldots\ldots (1)$$

calculations being made in pounds and in feet.

When we examine this formula we see that it may be put in another form. Find a point K, such that if all the mass of the pendulum were gathered there, its moment of inertia about s would be the same as at present; in fact, such that $\frac{W}{32 \cdot 2}$ the mass of the pendulum $\times$ s K^2 would be equal to I. The distance s K is called the radius of gyration of the pendulum (see Art. 142), and our rule now becomes,

$$T = 6\cdot 2832 \sqrt{\frac{s\,\text{K}^2}{g \cdot s\,\text{G}}} \quad \ldots\ldots\ldots\ldots (2)$$

where g is 32·2.

In the simple pendulum, s K and s G are equal, and if you make them equal you will find this to be the same rule which is given in Art. 177. However, in an ordinary pendulum, s K and s G are not equal, but s $\text{K}^2 \div$ s G is equal to some length such as s P, and our rule becomes

$$T = 6\cdot 2832 \sqrt{\frac{s\,P}{g}} \quad \ldots\ldots\ldots\ldots (3)$$

Evidently s P *is the length of the imaginary simple pendulum which would vibrate in the same time as our real pendulum.* The imaginary point P has been called the centre of oscillation, because when the pendulum is inverted and made to vibrate about an axis through P it vibrates in the same time as before.*

* To prove this it is necessary to return to equation (1). We know that I is equal to the moment of inertia of the body calculated as if all its mass existed at G, together with the moment of inertia of the body as it is at present, but calculated about an axis through G parallel to the present axis—that is

$$I = \frac{W}{g}\,s\,\text{G}^2 + \frac{W}{g}\,k^2,$$

where k is some length unknown to us just now, being the radius of gyration about the axis through the centre of gravity. Rule (1) becomes

$$T = 6\cdot 2832 \sqrt{\frac{\frac{W}{g}s\,\text{G}^2 + \frac{W}{g}k^2}{W \cdot s\,\text{G}}}$$

$$\text{or } T = 6\cdot 2832 \sqrt{\frac{s\,\text{G} + \frac{k^2}{s\,\text{G}}}{g}}$$

That is, the length of the simple pendulum which will vibrate in the same time is $s\,\text{G} + \frac{k^2}{s\,\text{G}}$, and we have already found it to be s P in equation (3), so that $\text{G}\,P = \frac{k^2}{s}$, or $\text{G}\,P \times s\,\text{G} = k^2$. But in the very same

186. Constraint of Spiral Spring.

186. *Examples.*—The bar of Fig. 129 with **two adjustable masses** may be hung at one end of a wire, the other

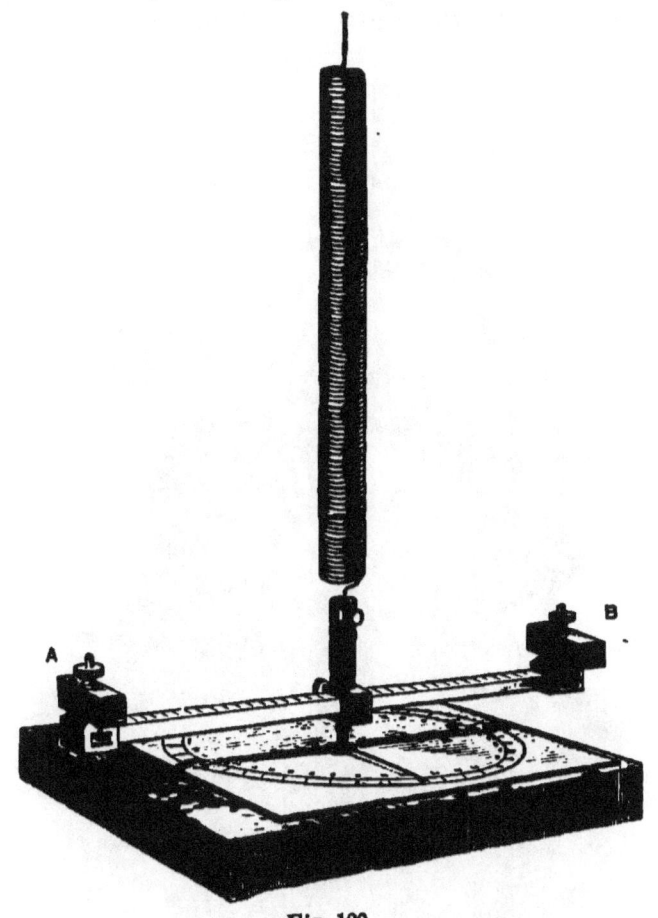

Fig. 129.

end of which is fixed to the ceiling. By twisting and untwisting the wire the bar will oscillate with a motion

way, if we considered the pendulum as vibrating about P, we should find the length of the equivalent simple pendulum to be greater than G P by an amount equal to $\frac{k^2}{G P}$, and we know that S G is equal to this amount, so that S P would as before be the length of the equivalent simple pendulum. *The axes of oscillation and suspension are therefore interchangeable.*

198 PRACTICAL MECHANICS. [Chap. XVIII.

which is much more nearly pure harmonic than that of the balance of a watch. My students experiment with such a bar; they can adjust the weights A and B at any

Fig. 130.

distance from the axis (there is an engraved scale on the bar), so that the moment of inertia can be varied. They can fasten the bar at the end of a wire, or they can use it as in Fig. 130, with a flat spiral spring, or as in Fig. 129, with a cylindric spiral spring; and the rate of its vibration gives one of the best ways of investigating the twisting moments of wires and such springs when strained through given angles.

In the case of a wire the twist always tends to bring the bar to its position of rest with *a moment which is proportional to the angle of displacement from this position*—it is this property which causes the motion to be pure harmonic. *This moment is also proportional to the fourth power of the diameter of the wire, and it becomes less as the length of the wire is increased.* By means of a circular scale and a pointer we can measure the extent of each swing, and this is found to decrease gradually, due to friction with the air and the internal friction or *viscosity* of the metal. The amount of diminution of swing gives us a means of determining the viscosity, and the apparatus can so easily be fitted up, that no person who wishes to understand the properties of materials can be excused from making these experiments.

If the length of the wire is l inches, its diameter d, and if N is its modulus of rigidity (see Table III.), then from Art. 92 we see that the moment with which the wire acts on the bar, when its angle is A from the position of rest, is

$$\frac{\pi}{32} N \frac{A}{l} d^4.$$

If the moment of inertia of the bar is I (we are neglecting the fact that the wire itself has some mass which has to be set in motion), then the moment, divided by I, is the angular acceleration, and using this quotient as denominator and A as numerator, extracting the square root, and multiplying by 6·2832 or 2π, by the general rule of Art. 184, we find the square of **the period of a complete oscillation** to be

$$T^2 = \frac{128 \pi I l}{N d^4}.$$

When motion is slow, the friction in fluids is proportional to the velocity, and any friction which follows this law is called fluid friction. A great many vibrating bodies tend to come to rest by the action of such friction as this; and it is found that if the friction is numerically f times the angular velocity, then the logarithm of the ratio of the length of one swing from the middle position, to the next swing in the same direction, is nearly equal to k times the periodic time. Hence, this logarithmic decrement, as it is called, is proportional to the friction co-efficient. If we observe twenty-one elongations on one side of the middle position, then one-twentieth of the logarithm of the first elongation divided by the last, is k times the periodic time of oscillation.

Exercise.—**Bifilar Suspension.**—In many measuring instruments a body is suspended by two thin wires nearly vertical. If the vertical length of each of these is l, the distance between their ends at the top, a, and at the bottom, b, and the weight of the body, w, it is easy to show that for a small angular displacement, A, the moment tending to bring the body to its position of rest is very nearly (neglecting tension of the wires themselves)

$$\tfrac{1}{4} \frac{ab}{l} w\text{A}.$$

Find the time of vibration of such a body when its moment of inertia is known.

Exercise.—A **magnet**, turning on a frictionless pivot at its centre of gravity, is subjected to a turning moment, HA, due to the earth's magnetic action, if it makes only a small angle, A, with its position of rest. Find the time of a vibration if the moment of inertia is known, and show that the square of the time of vibration of the magnet in different places is inversely proportional to H.

187. Stilling of Vibrations.—When a pure harmonic motion is represented on paper in the manner described in Art. 181, we have a curve of sines. The curve may be obtained by producing the lines B B′, C C′, &c., of Fig. 116, cutting them at right angles by equi-distant horizontal lines and joining the successive points of intersection so found. It may also be drawn by finding from a book of tables the sines of 0°, 10°, 20°, &c., and plotting 0 and sin 0°, 10 and sin 10°, 20 and sin 20°, &c., on a sheet of squared paper.

A *curve of sines* expresses the fact that, if d represents the displacement of a vibrating body from its middle position after an interval of t seconds since it was at the middle of its course, then

$$d = \text{A} \sin \text{B}\, t$$

where A is the greatest displacement of the body from its

Chap. XVIII.]   STILLING OF VIBRATIONS.   201

middle position. This displacement is usually called the *amplitude of the vibration*.

If T is the time of a complete vibration, it is easy to see that the equation is

$$d = A \sin \frac{2\pi}{T} t.$$

If we make the bob of a pendulum terminate below, in a tube which can act as a pencil holder, and in which a well-fitting pencil can slide freely, and if we move a sheet of paper at a uniform rate underneath this pencil at right angles to the direction of motion of the pencil, a curve of sines will be traced out, if the pendulum swings without friction. But in practice we always find that, what with the friction at the point of support, friction with the atmosphere, &c., a pendulum's swings get smaller and smaller, that is, **the amplitude of the vibration gets less and less as time goes on, until the pendulum at length comes to rest.**

This motion is not a pure harmonic motion, but within certain limits each swing may be regarded as very nearly a pure harmonic motion. Practical men who deal with oscillating bodies, such as pendulums, ships, tuning forks, magnetic needles, and suspended coils of wire, usually assume that the motion during each swing is a pure harmonic motion. The frictional resistance to motion of any ordinary vibrating body in a fluid medium, or of a magnetic needle vibrating near any body capable of conducting electricity, is almost always such that the quicker the motion the greater the friction (see Art. 244), that is, frictional resistance is proportional to speed, and in this case it is not difficult to show that instead of the law

$$d = A \sin \frac{2\pi}{T} t \quad \ldots\ldots\ldots\ldots\ldots (1),$$

we have the law

$$d = A \epsilon^{-kt} \sin \frac{2\pi}{T_1} t \quad \ldots\ldots\ldots (2).$$

That is, if the strength of the spring or other governor of vibration, and the character of the vibrating body are such that without friction the law would be (1), then, when the vibration is damped by frictional resistance of the above character, the law of the motion becomes that given by equation (2). Here k is a constant which depends on the character of the friction. Thus k is greater when a pendulum swings in water than when it swings in air. Also, T_1, the periodic time of the vibration, is no longer the same T as it was for undamped vibrations, and the relation between T and T_1 is

$$\frac{1}{T^2} = \frac{1}{T_1^2} + \frac{k^2}{4\pi^2} \quad \quad (3)$$

or if n is the number of undamped vibrations per second, and n_1 the number of damped vibrations per second, then

$$n^2 = n_1^2 + \frac{k^2}{4\pi^2} \quad \quad (4).$$

In order to get exact ideas on this subject of the **damping of vibrations**, the student ought to plot on squared paper a curve such as O A' B' C' D' E' F' G' H' I, Fig. 131, which corresponds with equation (2). Thus, let us suppose that a body undamped in its vibrations gets an impulse which sends it from its position of rest in such a way that its amplitude is 10 inches, and let the time of a complete oscillation be 1·6 second. Then the law of its motion would be

$$d = 10 \sin \frac{6 \cdot 2832}{1 \cdot 6} t,$$

or

$$d = 10 \sin 3 \cdot 927 t \quad \quad (5)$$

where d is in inches, t in seconds, and the angle $3 \cdot 927 \, t$ in radians (see GLOSSARY, Art. 230).*

If now the friction is such that $k = 0 \cdot 7$, we find from (3) that the time of a vibration is practically unchanged. Find therefore the original curve of sines by calculating the second column of the following table. The numbers of the first two columns plotted on squared paper would represent the undamped vibrations. But for damped vibrations the numbers of the second column have all to be multiplied by $\epsilon^{-0 \cdot 7t}$, and if we denote this multiplier by the letter x, we see that

$$x \text{ being } \epsilon^{-0 \cdot 7t}$$
$$\log x = -0 \cdot 7t \log \epsilon,$$

or $\log x = -0 \cdot 304 t.$

---

* We may write (5) in the form

$$d = 10 \sin 225 t.$$

In this case the angle $225 t$ is expressed in degrees.

I have calculated x for the various values of t, and placed the results in the third column. Multiplying, therefore, the respective numbers of the second and third columns together, we get the fourth column of numbers, and plotting the numbers of the first and fourth columns on squared paper, we find the curve which shows the nature of the damped vibrations.

t in seconds.	10 sin 3·927t, or (10 sin 225t, if angle is taken in degrees.	$\epsilon^{-0.7t}$	$\epsilon^{-0.7t}$ 10 sin 3·927t
0	0	1	0
0·2	7·07	·869	6·14
0·4	10	·756	7·56
0·6	7·07	·657	4·65
0·8	0	·571	0
1·0	−7·07	·497	−3·51
1·2	−10	·432	−4·32
1·4	−7·07	·375	−2·65
1·6	0	·326	0
2·0	10	·247	2·47
2·4	0	·186	0
2·8	−10	·141	−1·41
3·2	0	·106	0
3·6	10	·080	·8
4·0	0	·061	0
4·4	−10	·046	−·46
4·8	0	·035	0
5·2	10	·026	·26
5·6	0	·020	0
6·0	−10	·015	−·15
6·4	0	·011	0

Some students may find it as instructive to first draw a curve of sines, then draw the logarithmic curve, corresponding to column three, on the same sheet of squared paper, and multiply the ordinate of one curve by that of the other to get the ordinate of the real curve which exhibits the damped vibrational motion. This is what has been done in Fig. 131; O A B C D E F G H I is the curve of sines, L P Q is the logarithmic curve,

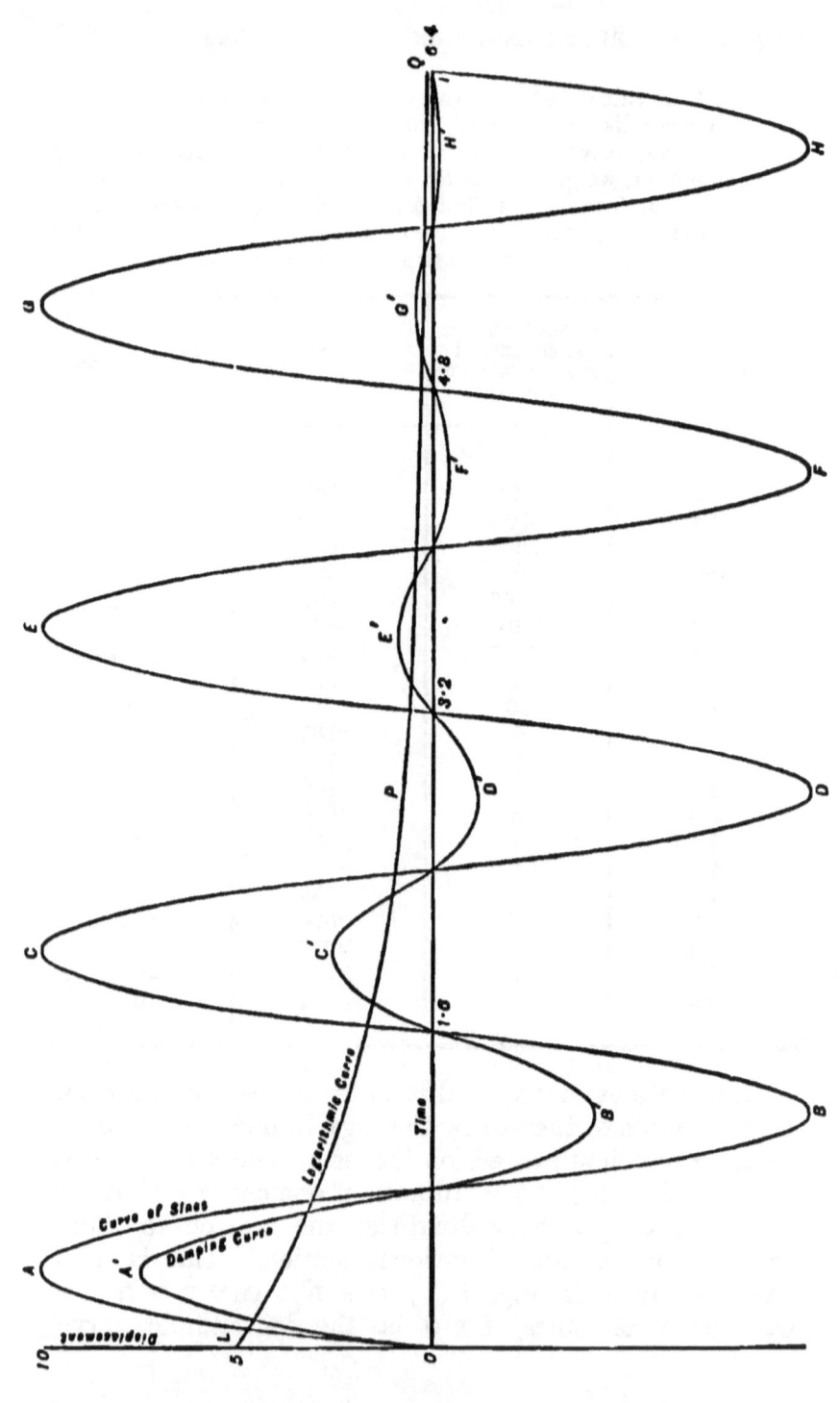

showing how rapidly the amplitude of the vibration diminishes, and O A' B' C' D' E' F' G' H' I is the curve which represents the actual motion of the vibrating body. In this figure the logarithmic curve is drawn to such a scale as seemed convenient for showing its properties distinctly. It would be very easy to dilate on the nature of the resulting curve O A' B' &c., but this book is written to help students who are earnest enough to calculate the above numbers and plot the curve, and when they perform these operations they will have very clear notions about the motion we have been investigating, so I think it would be mistaken kindness on my part to describe it further.

Exercise.—A heavy disc, suspended by a wire, vibrates in each of a number of fluid media, its periodic time of vibration in all being sensibly the same, or 1·5 second. The ratio of the amplitudes of two successive swings in one direction being 0·9 in one fluid (that is, the second swing being only nine-tenths of the first, and the third being only nine-tenths of the second, and so on) and 0·8 in another fluid, and 0·7 in another, what numbers will express the relative viscosities of these fluids?

Here we have
$$0.9 = \epsilon^{-1.5k}$$ for the first fluid,

so that $-\log 0.9 = 1.5k \log \epsilon$,

or $k = \dfrac{-\log 0.9}{1.5 \log \epsilon}$,

that is $k = 0.07$.

In the same way $k = 0.15$ and $k = 0.24$ for the other fluids, and hence 7, 15, and 24 are the required numbers expressing the relative viscosities, as measured by the vibrating disc method.

A very slowly swinging disc and pointer will enable you to plot the complete curve from actual observations.

CHAPTER XIX.

THE EFFECT OF A BLOW.

188.* Have you ever considered how it is that a blow of a hand hammer will indent a steel surface, whilst a steady force applied to the same hammer-head would require to be very great to produce any indentation? The pressure between the hammer and the steel is very great, and it must be all the greater because the time of contact is very short. Indeed, if a hammer weighs 2 lbs., so that its mass is $2 \div 32\cdot 2$ or $\cdot 0621$, and if, just before touching the steel, its velocity was 10 feet per second, then you know from the principles of mechanics that its momentum was $10 \times \cdot 0621$, or $\cdot 621$. Now, if one-ten-thousandth of a second elapses from the time of actual contact until the hammer's motion is destroyed—that is, until the elasticity of the steel is just about to send the hammer back again a little—the momentum $\cdot 621$ is destroyed in $\cdot 0001$ second, hence the average force acting between hammer and steel during this short time must have been $\cdot 621 \div \cdot 0001$, or 6210 lbs. It is certain, however, that this average force is less than what the force actually was for some very small portion of the time. You will observe then that we cannot tell the **average force of an impact** unless we know two things—first, the **momentum**, and second, the **time in which it was destroyed**. Now the duration of an impact depends greatly upon the nature of the objects which strike one another, and we see that the average force of a blow is less as the time is greater.

* In Art. 63 I have touched on this subject from the **strain-energy** point of view. It might have confused the student to treat the subject from two different points of view in one place. This is my reason for the separation.

Sometimes, instead of a great force acting for a very short time, what we require is a smaller force acting for a longer time. For instance, when cutting wood we obtain this result by using a wooden mallet and a chisel with a long wooden handle, because the force required to make the chisel enter the wood is not very great, and we wish this force to act for some time, so that much wood may be cut at one blow. In *chipping*, we have the time short, because considerable force is required to cause the chisel to enter metal. The duration of an impact depends on the shape of the bodies and their masses, and on the rigidity of their materials.

Why is it that in driving a nail into wood your blows seem to be of no effect unless the wood is thick and rigid or unless it is backed up by a piece of metal or stone? It is because the wood yields quite readily, and so prevents the hammer losing its momentum rapidly. There are few subjects in which people are so apt to have erroneous ideas as in this one of impact. Thus a man will speak of the *force* produced by a weight falling through a height without having any idea of the *time* during which the motion of that weight is being stopped, in fact, without considering what time the weight is allowed for delivering up its energy. Now, a little consideration will show that the mean force of the blow will be quite different according as the weight falls on a long and yielding bar or on a short and more rigid one. If we could imagine bodies to be formed of perfectly unyielding materials, then the slightest jar of one against the other would produce an infinitely great pressure between them, and in the blow produced by a falling body there may be every gradation from exceedingly great pressures to very small ones, depending on the yielding power of the body that is struck. Everybody is acquainted with the sensation produced by suddenly placing one's foot on a level floor when one was preparing for a step downwards. The downward momentum of the body is suddenly destroyed, and great pressures have to act in all the bones of the body. Carriages are hung on springs for the purpose of preventing their losing or gaining momentum with too great rapidity when the carriage wheels pass over obstacles. When we are sitting on a hard seat in a third class railway compartment, and the carriage gets a slight jerk upwards, momentum is given much too rapidly to our bodies for perfect comfort, and to sit on cushioned seats is preferable. A cannon ball is safely, because comparatively slowly, stopped by sand-bags or bales of cotton.

190. *Example.*—A pile driver of 300 lbs. falls through a height of 20 feet, and is stopped during 0·1 second, what average force does it exert upon the pile? A body which has fallen freely through a height of 20 feet has acquired a velocity equal to the square root of 64.4×20, or 35·89 feet per second. Its momentum is $300 \div 32.2 \times 35.89$, or 334·4, and this divided by 0·1 gives 3,344 lbs., the answer. From the instant when the motion of the driver ceases to diminish, the force exerted by it is its own weight. The average force of friction in pounds between the pile and the ground multiplied by the distance in feet through which the pile descends during the stroke is equal to 300×20, or 6,000 foot-pounds, if we neglect the loss due to vibrations of the body and the ground in the neighbourhood, and if we also ignore the fact that the weight really descends a little farther than 20 feet.

190. Suppose a body A to strike another B, and that we can neglect the actions of outside bodies upon them both. If A loses momentum, B must gain the same amount because their mutual pressures are equal and opposite during the time of impact. It is our knowledge of this fact that enables us to calculate the motions of bodies after they strike one another. Again, for the same reason, if from any internal cause the parts of a body separate from one another, either violently or gently, the **total momentum** remains as it was, it is only the relative momentum which alters.

Hence, when a shell bursts in the air, some parts move in the same direction more rapidly than before, but others less rapidly; one part may double its velocity and another may drop nearly vertically, its forward motion being stopped, but, on the whole, the total forward momentum is what it was originally.*

191. *Examples.*—If a cannon were perfectly free to move backward when the shot leaves it, the backward momentum of the cannon would be exactly equal to the forward momentum of the shot. Thus, if a shot of 20 lbs. leaves a cannon whose weight is 2,240 lbs. with a velocity of 1,000 feet per second, the velocity of the cannon backward would be $1,000 \times 20 \div 2,240$, or about 9 feet per second, neglecting the fact that the gases leave the gun also with a certain momentum. When a ship fires her broadside, each gun runs back, communicating, as it is stopped, its momentum to the ship, which heels over in consequence. A gun firing the above shot of 20 lbs. directly backwards from a ship whose total weight is 600 tons gives to the ship so much momentum that its speed is increased $9 \div 600$, or ·015 foot per second. We see, then, that a ship might propel herself by means of her guns. The steamship *Waterwitch* had powerful steam pumps, wherewith she brought a great quantity of water in nearly vertically, and sent it out backwards on the two sides below water level. The momentum given to the water backwards was equal to the momentum given in the other direction to the ship. It is on this principle that Hero's steam-engine and Barker's mill work, the momentum given to jets of fluid passing out of certain pipes being equal to the momentum given in an opposite direction to the pipe from which the fluid passed. In all such cases the propelling

---

* Of course the kinetic energies of the parts of the shell added together are greater than they were before the shell burst; we are now merely speaking of the momentum. The total momentum of two equal bodies going in opposite directions with the same velocity is nothing, whereas their total kinetic energy is double that of one of them.

o

force in pounds is numerically equal to the momentum of the fluid which passes out in a second. Thus, if from a vessel moving with a velocity of 14 feet per second water comes through orifices of 4 square feet in area with a velocity (relative to the orifices) of 20 feet per second, then the quantity passing out in one second is 4×20, or 80 cubic feet, that is $80 \times 62 \cdot 3$, or 4,984 lbs. Now recollecting that this water was first brought in and is now sent out, what is the velocity which we have really impressed upon it in the process? At the beginning it was motionless with respect to the sea; it now has a velocity of $20 - 14$, or 6 feet per second with respect to the sea, so that the momentum given to it is its mass $4,984 \div 32 \cdot 2$ multiplied by 6, or $928 \cdot 7$: hence, as this momentum is given every second, $928 \cdot 7$ lbs. is the propelling force exerted on the ship. In one second the ship moves through 14 feet, so that the useful mechanical work done is $14 \times 928 \cdot 7$, or 13,002 foot-pounds. We have given to 4,984 lbs. of water a velocity of 6 feet per second, the kinetic energy of this water is wasted, and this kinetic energy is $\frac{1}{2}$ of $4,984 \div 32 \cdot 2 \times 6 \times 6$, or 2,786 foot-pounds. In fact, we have altogether spent 15,788 foot-pounds, and 13,002 of this have been usefully employed, so that the efficiency of the method is $13002 \div 15788$, or $0 \cdot 824$, or $82 \cdot 4$ per cent. As a matter of fact, however, the friction in pumps and pipes usually causes a third of the indicated horse-power of the engine to be wasted, so that the true efficiency of this method of propulsion is $\frac{2}{3}$ of the above, or $0 \cdot 549$, or $54 \cdot 9$ per cent., neglecting the efficiency of the cylinder of the engine itself. You will remember a fact which has come in casually here—if the water leaves any turbine, water-wheel, or any propeller of a vessel with a velocity relative to the still water into which it passes, or if it has any other form of energy, this energy has been wasted. After studying the above calculation, you will have no great difficulty in understanding how we find the horse-power given out by turbines.

192. By calculation you will find that, when two free and elastic bodies strike, the momentum communicated from one to the other is their relative velocity multiplied by the product of their masses and divided by the sum of their masses, and this quotient divided by the time of impact gives the mean pressure. This pressure acts equally on both, of course, but it may not hurt both equally. If the bodies are surrounded by water like ships, they can no longer be regarded as free bodies, and it is not easy to tell you in a few words how much mass you must add to the bodies to represent the mass of the water, which has also to undergo change of motion. In the case of a ship, the mass of water to be moved broadside on is much greater than when the ship is struck stem on.

193. A body falling into a liquid sets it in motion, and this motion appears at distant places more and more nearly *instantaneously* as the liquid becomes more and more incompressible. The nature of this motion is known to us if we know the velocity of yielding at the place of contact, and from this the total momentum given to the liquid. This represents a very considerable pressure applied at the place of contact, and this pressure becomes greater as the velocity of the body before it touches the liquid increases. Hence a cannon ball fired at sea rebounds from the water as from a rigid body. Hence also a man diving unskilfully, as he falls prone on the water, gets a very unpleasant shock, whereas a skilful diver enters in such a way as to make the momentum of the moving water as small as possible, and to make the creation of this momentum gradual.

194. Let us consider what takes place when two free ivory balls come together. There is a certain instant after they first touch when they move together just as if they were composed of soft clay—then they act on each other with their greatest pressure; they are in their most strained condition, and supposing no loss by internal friction, the strain in the balls represents an amount of stored-up energy (see Arts. 62, 63) equal to the kinetic energy which the bodies have lost. It is very important to remember this fact, that if bodies are to return to their old states after the collision, we must suppose that during the collision there is a storage of kinetic energy in the form of strain. All the kinetic energy will not be given out again, nor can we say that it is all stored, because there is a sort of internal friction causing part of the strain energy to be converted into heat when any change occurs. Now, if the whole of the stored-up energy is confined to one portion of the body, the strain may be too great. Thus, a steel rod 1 square inch in section, 1 foot long, will store up 167 foot-pounds of strain energy in its stretched condition before it breaks. For suppose breaking stress to be 100,000 lbs. per square inch. This will occur when there is a lengthening of ·0033 foot, so that the energy stored up is the work done by a force whose average amount is 50,000 lbs. acting through ·0033 foot, or 167 foot-pounds. If 2 feet of the same rod stored up this same amount of energy, there would only be 83 foot-pounds in each foot of its length, and it is easy to see that the stress is no longer the breaking stress of 100,000 lbs. per square inch, but only 70,700 lbs. per square inch. As we store the same amount of energy in smaller and smaller portions of a body, it is evident that we must approach a condition of fracture.

195. We see, then, that at the place where contact occurs, two bodies, A and B, are strained; but if A is of some very elastic material, such as tempered steel, the strain energy is conveyed very rapidly to every part of the body, whereas if B is a feebly elastic body, the strain accumulates at one place, leaving the rest of the body unstrained, whilst at this place the strain may produce fracture. This slowness to communicate strain to the rest of the body may also

be produced by the shape of the body. For instance, a rod struck sidewise or a thin plate struck in the middle does not so immediately communicate its strain to the remote parts as a rod struck endwise. Again, the nature of the parts of A and B in contact may be such that not only does the strain energy leave this part of A rapidly, but immediately in the neighbourhood of the place of contact there is a greater capacity to bear strain energy without rupture than is the case with B. Thus, when ship A rams the broadside of ship B, the side of B is bent inwards and the strain energy produced is accumulated near the place of contact till fracture occurs, whereas, not only is A's stem able to transmit to all parts of A with great rapidity the strain energy which must be stored up in the whole mass, but at the stem itself the material of A is capable of withstanding greater stresses than the material of B's side. Suppose, however, that A's stem is not of steel, still B's iron or wooden side will be perforated if A has enough velocity; A's stem may also be damaged in the impact in such a case.

196. A candle may be fired, it is said, **through a thin deal board** with very little injury to its shape, and the usual explanation of this phenomenon given in books is that the candle has not time to get broken. This explanation is not satisfactory; it is a little too vague. If we had the board in rapid motion, and striking the candle in the same relative position, the candle having previously been at rest, would the candle perforate the board? There cannot be any doubt that it would. Hence it is not the body struck which must in every case get hurt; the pressure on one is equal to that on the other. Suppose ship A rushes at ship B when B is broadside on, and rams her, B will probably be sunk, even if she is a much larger and better ship than A. But suppose that B is able to meet her adversary stem to stem, if they are equally strong they will equally injure one another, and if B is the stronger A will suffer the most. This case differs very much from that of the candle, because we can assume greater strength even for slowly applied pressures from stem to stern of the ship A than from side to side of B; whereas the strength of the candle for slowly applied pressures cannot be compared with that of the wood which it punches from the board. What is meant by the usual explanation, "The candle has not *time* to get deformed?" Why has not the soft candle time to get broken, and yet the wood has time to get torn asunder? The fact is, the wood, if it were slowly pressed, would communicate its strain energy to every part of the board and its supports; but this communication takes an appreciable interval of time, however suddenly the pressure may be applied, or however great it may be. As the strain energy is rapidly produced it becomes accumulated near the place of contact to such an extent as to produce fracture of the wood. Now the point of the candle is subjected to the same pressure as the wood, and begins to get spoiled in shape—that is, it is compressed—and this compression produces a lateral spreading. In the meantime, however, the compressive strain energy is communicated very rapidly backwards along the candle, and the spreading and spoiling goes on along its entire length, but is small at any point, since it is distributed over the whole mass. Practically therefore the spoiling occurs only at the point of the candle since time is needed for fracture of the material.

197. **An earthquake,** when it acts on a house, usually tends to move it through a distance of probably a very small fraction of an inch, but it does this in a very short time—that is, the house gets a considerable velocity. The mass of the house multiplied by the greatest velocity and divided by the short time during which the momentum is being communicated, gives the pressure which the foundations of the house are subjected to. Now, when the foundations are not very rigidly connected with the ground, **the time of communication of the momentum is lengthened,**

and the pressure is consequently diminished. This is the usual Japanese plan of providing for earthquake effects. Unfortunately, the very means taken to diminish the pressure on the foundations also diminishes their capability of withstanding pressures, and it has not yet been decided what sort of a house is best fitted to withstand destructive earthquakes. Want of˙rigidity, combined with strength or toughness in the materials, and especially the quality of internal friction in the materials, so that vibrations may rapidly die away—these are the qualities needed. They are found in steel, wrought iron, and wood, and especially in wicker-work, in a less degree in cast iron and in brick or stone set in cement, and less still in brick and stone set in bad mortar.

198. When you in some way understand the possibility of a candle perforating a board, you will be able to comprehend how sand, when blown in air against tempered steel, is able to abrade it; how the emery wheel and grindstone going at great velocities are able to cut into hard metals, and how in California a jet of water going with very great velocity is used for mining purposes instead of iron tools.

199. **Quasi-Rigidity produced by Rapid Motion.—** A top when not spinning can with difficulty be balanced on its point, and if left to itself it almost instantly falls; whereas when it is spinning the effect of slightly tilting it out of the perpendicular is not to make it fall, but to make it take a slow precessional motion.

There is a piece of apparatus called a gyrostat, which is, in a more or less perfect form, to be found in every mechanical laboratory, and the student ought to experiment for himself with this apparatus on the curious effects of quasi-rigidity which manifest themselves in tops and other spinning bodies. If he has a slight acquaintance with astronomy he will be interested in tracing the connection between the behaviour of a tilted top and the precession of the earth's equinoxes.

When a circular sheet of drawing paper is mounted

like a very thin grindstone on an axis, and is gradually made to rotate rapidly, it is found to have become quite rigid, that is, it greatly resists bending as if it were made of steel. In the same way a long loop of rope, hanging round a high pulley, which gives it a quick motion, takes a certain form which it is very difficult to alter, as may be shown by striking it with the hand or with a stick: it resembles more a rigid rod than a flexible rope.

Again, in the well-known lecture-experiments on smoke rings, we see that these little whirlpools of air have many properties in common with elastic solid bodies on account of the partial rigidity which is due to their rapid motion.

It would be beyond the scope of an elementary book like this to explain these curious phenomena, and I merely direct the attention of students to these instances in order to incite them to make experiments, and to seek for the explanation of what they observe.

200. Motion Produced by a Blow.—When a body subjected to a blow is quite free to move in any way, unless the blow acts through its centre of gravity, the body will not merely move as a whole, but it will revolve. When the blow acts in a direction through the centre of gravity there is no rotation produced. It is usual in such a case to consider the motion of the centre of gravity of the body, and the motion of the body about an axis through its centre of gravity, for it is known that any motion whatsoever of the body, consists of a combination of two such motions. It is found that the kinetic energy communicated to a body by means of a blow is best calculated in the following way, if we know the nature of the motion:—
First, find the kinetic energy, as if every portion of the body had the motion of the centre of gravity. Secondly, find the kinetic energy of rotation as if the axis of rotation through the centre of gravity were fixed in space. Add these two results together. We may

regard from another point of view the instantaneous motion of a body when it is struck, namely, as a rotation about some axis which does not itself move. This is only the case for an instant; immediately afterwards we must regard the body as moving about another fixed axis. If a body is hinged so that it can only move about a fixed axis, it is always possible to find the point at which the body may be struck, and the direction of the blow, which will tend to produce an instantaneous rotation about this particular axis, and therefore to produce no pressure at the hinge. Thus the *ballistic pendulum* of Fig. 132 is always struck in such a way, and the point in which it is struck is called the "**centre of percussion.**" An easy way to find the centre of percussion is as follows:—Make the body vibrate like a pendulum, about its axis of suspension, under the action of gravity. Now find the *length of the equivalent simple pendulum*. This is the distance of the centre of percussion from the axis. In a tilt hammer all blows ought to be delivered from this centre of percussion if we wish to have no pressure on the bearings. A cricket-bat or a rod of iron tingles the hand when we strike a blow with it, unless we

Fig. 132.

happen to strike at the centre of percussion. For a rod of iron free to move about one end, the centre of percussion is at two-thirds of the way towards the other end. (See Art. 185.)

201. The **Ballistic Pendulum** of Fig. 132 is a contrivance which enables us to measure the velocity of a bullet. It consists of a mass of wood, A, forming part of a pendulum. The bullet is fired into it, and the wood swings backwards in consequence. The bullet is fired into that part of it which will cause no jar to be given to the pivot B. The momentum existing in the bullet before it enters the wood belongs now to the whole mass of which it becomes a part. C is a silk ribbon which is pulled through a moderately tight hole by the swing of the pendulum, and the length of ribbon pulled through is found to be proportional to the momentum of the bullet before entering the wood.

CHAPTER XX.

THE BALANCING OF MACHINES.

202. If a wheel is fixed eccentrically on its shaft, or if to a shaft there is attached any object whose centre of gravity is not exactly in the axis when the shaft rotates, **centrifugal force causes pressures on the bearings of the shaft** which are always in the direction of the centre of gravity of the rotating mass. In this case there is said to be a want of balance. If you wish to observe the effect produced by such want of balance, mount an axle to which a wheel is keyed on any support which is not very firm; fix a small weight on one of the arms of the wheel, and rotate it rapidly. You will find that, even if the weight is small, surprising effects are produced, and show themselves in a shaking of the supports; and the evil effects are four times as great at 200 revolutions per minute as at 100 revolutions per minute. **Centrifugal force is proportional to the mass of a rotating part multiplied by the distance of its centre of gravity from the axis of rotation, multiplied by the square of the number of revolutions per minute.**

203. If a number of bodies are attached to a shaft and are whirling round with it, each of them exerts a force on the shaft which can be calculated, and the resultant effect on the two bearings may easily be determined. If the axis of rotation passes through the centre of gravity of all the rotating parts, the pressure on one bearing is equal and opposite to the pressure on the other, and by properly placing the masses the pressure on the bearings may be reduced to nothing. Thus it is evident that when two masses are directly opposite to one another on a shaft, their centrifugal forces may be made to balance one another. When not opposite they cannot be made to balance, but two masses may balance one which is directly opposed to the resultant force of the two. When there is neither pressure on the bearings nor tendency to change the direction of the axis, it is said to be the *permanent axis* of the rotating masses. **All axes of rotation in machines ought to be permanent axes.** When this is the case in a locomotive engine, and it is suspended like a pendulum bob, by two ropes fastened to the driving shaft, and made to work, there are no visible oscillations.

204. The balancing of a machine consists in adding masses in such positions, or re-arranging the positions of the existing masses, so that the centrifugal forces due to their rotation are just able to balance the otherwise unbalanced forces which act on the various shafts. The student will find that the study of one problem in balancing will make him familiar enough with the method of calculation for its application to almost any other case which is likely to occur in practice. The most usual case for the student to take up is that of the locomotive engine, because want of balance in the locomotive is capable of producing very serious effects indeed.

205. *Example*.—It has been shown by experiment that the application of suitable balance weights is attended by a sensible reduction of resistance on railways at high speeds. Locomotive engines unbalanced cannot attain as high speeds as when balanced, with the same consumption of fuel. There are two separate sets of unbalanced forces acting on the crank shaft of a locomotive. (1.) The centrifugal force of the crank, crank-pin, and as much of the connecting rod as may be supposed, roughly, to follow the path of the crank-pin (say one-half of it). The mass or weight of each of these multiplied by the distance of its centre of gravity from the axis, divided by the length of the crank, gives the mass which, on the crank-pin, would produce the same centrifugal force. Let this weight be called w lbs. In designing engines it is better to trace out the curved path of the centre of gravity of the connecting rod, and calculate exactly the nature of the pressure which its motion produces on the axis, but for rough calculation we consider half the rod to act as if collected at the crank-pin, the other half to be moving with the piston.

At the end of the stroke, when the horizontal component of the centrifugal force is greatest and the vertical component vanishes, the horizontal pressure on the axle caused by the centrifugal force is

$$w \text{ R } n^2 \div 2{,}937,$$

R being the length of the crank in feet and n the number of revolutions per minute. (See GLOSSARY, Art. 234.)

(2.) We have the force due to the momentum of the reciprocating mass, including piston, piston-rod, slide, and the second half of the connecting rod. The loss of momentum is most rapid just at the end of the stroke; and as loss of momentum per second is what we call force, the force acting on the axle at the end of the stroke due to this cause is easily found and proves to be

$$\text{W R } n^2 \div 2{,}937,$$

where W is the weight of the total reciprocating mass.

Now a weight w_1, or weights whose sum is w_1, may be placed on the driving-wheel or wheels at a distance r from the axis, such that the centrifugal force of w_1 may be equal to the sum of the above forces. This leads to

$$w_1 r = \text{W R} + w \text{ R};$$

and if we assume any distance, r, we can calculate the balance weight or weights, w_1.

206. Now, for the axis to be permanent in inside-cylinder engines, w_1 must be divided into two parts, one for each wheel, *inversely proportional to the distances of the wheels from the crank*. For outside-cylinder engines we get balance weights for the two wheels whose *difference* is w_1, and they are, as before, inversely proportional to the distances from the wheels to the crank in question. Hence, a consideration of each cylinder gives two balance weights, one usually much smaller than the other. As the cranks are at right angles, the balance weights ought to be 90° apart on each wheel. Instead of using these two we can use one weight placed between their positions, so that its centrifugal force is the resultant of theirs. Thus, if we found 20 lbs. and 6 lbs. for the two placed at the same distance from the axis but 90° apart, make O A equal 20, and O B, at right angles to O A, equal to 6 according to any scale; complete the parallelogram, and O C represents on the same scale the weight which will replace them. It ought to be placed at just the same distance from the axis as they were supposed to be placed, and in position it makes the angle A O C with the larger weight. In this case it will be found that 20·88 lbs. placed 18·3° from the position which the weight of 20 lbs. might have occupied will be required to replace the two.

207. It often happens in outside-cylinder engines that the distance from one wheel, or rather from the centre of gravity of a balance weight, to the crank, is so little that the corresponding weight for the other wheel is very small and may even be neglected. In inside-cylinder engines it will be found that, whereas the cranks are at right angles to one another, the balance weights on the two wheels on the opposite side of the axis to the cranks are often only 50° apart. In inside-cylinder engines with coupled wheels, the *outside coupling rods and cranks are usually made to balance the inside moving parts*. These engines work very smoothly indeed. Outside-cylinder engines with coupled wheels are very unstable, from the use of small wheels requiring very rapid revolution of the crank axle; from the cylinders being farther apart than usual, so that the coupling rods may have room, and from the number of reciprocating parts being increased. The conditions seem to admit of no remedy for these defects. The balance weight ought to be distributed over two or three of the spaces of the wheel, that the tire may not be unduly strained. The reciprocating parts of engines ought to be as light as possible, and the width of cylinders as nearly as possible equal to that of the wheels.

208. We have, then, the following easy, approximately correct rules for locomotives:—If R is length of crank, r the distance of centres of gravity of every balance weight from centres of wheels, e the distance apart of the centre lines of cylinders, d the distance apart of the wheels or centres of gravity of the balance weights, w the total weight of crank (referred to the pin, see Art. 205), pin, connecting rod, piston, slide, and piston-rod, A the angle which the position of centre of gravity of balance weight makes with near crank:

(1.) Inside-cylinder engines with uncoupled wheels.

$$\text{Each balance weight} = \frac{w\,\text{R}}{2\,d\,r}\sqrt{2\,d^2 + 2\,e^2},$$

$$\tan.\ \text{A} = \frac{d-e}{d+e};$$

so that A is easily obtained from a book of tables.

(2.) Outside-cylinder single engines with uncoupled wheels.

$$\text{Each balance weight} = \frac{w\,\text{R}}{r},$$

$$\text{A} = 180°;$$

so that in this case the balance weight is placed exactly opposite to the crank.

(3.) Inside-cylinder engines with wheels coupled. Find by rule (1) if the weight of the coupling rods, &c., is too great. If so, let counter weights equal to the difference be placed opposite the outside cranks. If too small, the difference must be made up with balance weights, as in rule (1). The positions of the outside cranks are found by rule (1).

(4.) Outside-cylinder coupled engines. Find revolving weight of coupling rods, &c., for each wheel. Also find sum of the weight of the piston, rod, slide, and half connecting rod. Divide this latter among the wheels, adding the given revolving weight already on them. Let this be used on each wheel according to rule (2).

CHAPTER XXI.

GLOSSARY.

209. Introductory.—At the outset I stated that my readers must have a previous acquaintance with the elementary principles of mechanics; but, inasmuch as I am afraid that some may be deceived as to the exactness of their notions, I think it necessary to give a chapter in explanation of such fundamental facts and definitions as are most likely to be misunderstood. This chapter I call a Glossary, for although very different from any glossary I am acquainted with, it serves exactly the purpose of one.

210. Vertical Line.—A line showing the direction in which the force of gravity acts. It is a line at right angles to the surface of still water or mercury.

211. Level Surface.—A surface like that of a still lake, everywhere at the same level and everywhere at right angles to the force of gravity. It is not a plane surface.

212. Curvature.—For any curve you can find by trial at any part, what circle will best coincide with the curve just there? The radius of this circle is called the *radius of curvature* at the place. But since we say, for instance, that a railway line curves much, when we mean that the radius is small, the name *curvature* is always given to the reciprocal of the radius. Thus, if the radius is 8 feet, we say that the curvature is $\frac{1}{8}$th. If at another place the curvature is $\frac{1}{9}$th, the change of curvature in going from one place to another is the difference between these two fractions.

213. Mass.—Mass is the quantity of matter in a body. The weight of a body depends on whether it is at London or Paris, or elsewhere. A body which weighs a pound, as indicated by a spring balance, in London would weigh 0·46 lb. on the planet Mercury, would weigh 2·67 lbs. on the planet Jupiter, and would weigh hardly anything near the centre of the earth, and yet it would always have the same mass. The weight of a body in pounds at London, divided by 32·2, is a number which we use to denote the mass of the body. The weight of the same body in Spitzbergen, divided by 32·25, would give the same number. The weight of the same body at Trinidad, divided by 32·09, would give the same number. The weight of the same body on the planet Jupiter, divided by 85·9, would give the same number. These divisors are experimentally determined; they represent the acceleration to the velocity of a falling body, due to the action of gravity at the various places. Divide, then, the weight of a body in the British Islands by 32·2, and you get the mass measured in a way which is very convenient for a great number of calculations.

214. **Velocity** is the speed with which a body moves. Find the time in seconds taken by a body to traverse a certain distance measured in feet. This distance divided by the time is called the average velocity. Thus, if a railway train moves through 200 feet in 4 seconds, its average velocity during this time is 200 ÷ 4, or 50 feet per second. If we find, with careful measuring instruments, that it moves through 20 feet in ·4 second, or through 2 feet in ·04 second, the velocity is 20 ÷ ·4, or 2 ÷ ·04, or 50 feet per second. It is important to remember that the velocity may be always changing during an interval of time, and therefore cannot be said, at any instant, to be equal to the average velocity. To get the velocity at any instant, we must make very exact measurements of the time taken to pass over a very short distance, and even this will only give us the average velocity during this short time. But if we make a number of measurements, using shorter and shorter periods of time, the average velocity becomes

more and more nearly the velocity which we want. Thus, at 10 o'clock, a man in a railway train making a careful measurement finds that the train passed over 200 feet in the last 4 seconds. He finds the average speed for 4 seconds previous to 10 o'clock to be 200 ÷ 4, or 50 feet per second. Another man finds that it passed over 100·4 feet in the two seconds before 10 o'clock, and finds during these two seconds an average velocity of 100·4 ÷ 2, or 50·2 feet per second. Another man finds 50·25 feet passed over in one second previous to 10 o'clock, which shows a velocity of 50·25 feet per second. Another man finds 25·132 feet passed over in half a second before 10 o'clock, and finds 25·132 ÷ 0·5, or 50·264 feet per second. Another man finds 12·567 feet in a quarter second before 10 o'clock, and his observation gives 50·268 feet per second, and so on. It is evident that the values given by these various observations are approaching the real value of the velocity at 10 o'clock.

Tabulating these results, I have :—

Intervals of Time in Seconds before 10 o'clock.	Average Velocity in Feet per Second.
4	50·00
2	50·20
1	50·25
$\frac{1}{2}$	50·264
$\frac{1}{4}$	50·268

Plot the two sets of numbers on squared paper, and draw a curve through the points so found. Produce the curve, and you have the means of finding the average velocity for an infinitely small interval of time before 10 o'clock. This is the required velocity. I need hardly say that few practical purposes require us to observe with so much accuracy as this the velocity of a body whose velocity is changing.

215. Acceleration.—This is the rate of change of
P

the velocity of a body. Thus, it is known that the velocity of a body falling freely

At the end of one second is 32·2 feet per second.
    „   „   two seconds is 64·4   „   „
    „   „   three „   96·6   „   „
    „   „   four „   128·8   „   „

and we see that there is an *increase* to the velocity of 32·2 every second. The acceleration in this case is always of the same amount—hence we call it *uniform* acceleration, and say it is 32·2 feet per second per second.

If a force *continues* acting on a body, its effect is to produce a uniform acceleration in the direction in which it acts. You may experiment with Attwood's machine, or simply use, as *the body acted on*, a small carriage running very freely on a very smooth level

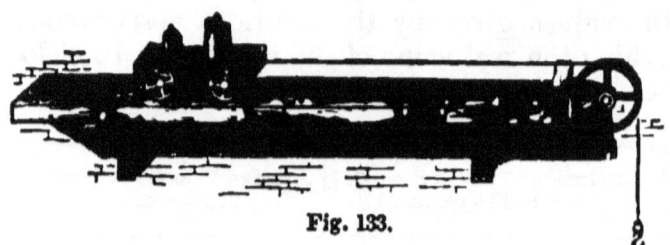

Fig. 133.

table; and *the force acting*, the pull in a string passing over a pulley on the edge of the table, and having weights in a scale-pan at its end, Fig. 133. Here, however, friction will be much greater than in Attwood's machine. You will prove the following rule to be true. If a force of 2 lbs. acts on a body whose weight is 50 lbs. at London (the 50 lbs. include the weight of everything which is set in motion, so that if you use a little weight of 2 lbs. for the purpose of exerting the force, remember that this little weight of 2 lbs. is included in the 50 lbs.), then the acceleration or increase of the velocity every second is equal to the force divided by the whole mass moved. In this case the mass is $50 \div 32\cdot 2$, or $1\cdot 553$, so that we have $2 \div 1\cdot 553$, or $1\cdot 223$ foot per second per second as the acceleration. Thus, if the body started from rest, the velocity would be $1\cdot 223 \times 5$, or $6\cdot 115$ feet per second at the end of 5 seconds. And now comes the question, how far will the body move from rest in five seconds? Evidently its average velocity during this time is half its terminal velocity, or $3\cdot 0575$, so that $5 \times 3\cdot 0575$, or $15\cdot 3875$ feet is the distance. It is evident that to get the space passed over we have multiplied half the acceleration by the square of the time.

When a body falls freely, its own weight is acting on its own mass. For instance, say the weight is 2 lbs., then the mass is $2 \div 32\cdot 2$, and weight divided by mass is acceleration, which we find to be in every case $32\cdot 2$ feet per second per second at London. The velocity at the end of any number of seconds is $32\cdot 2$ multiplied by this number; and the space fallen through in any number of seconds is half $32\cdot 2$, or $16\cdot 1$ multiplied by the square of the number of seconds. You can check these rules by the rules given you for potential and kinetic energy, and you will find them quite consistent with one another. Get to thoroughly understand the laws of energy first, and you will find that these other matters can be reasoned out quite easily afterwards.

216. Momentum.—If a body's weight is 2 lbs., its mass is $2 \div 32\cdot 2$, or $\cdot 0621$. Now, if the body is moving with a velocity of 20 feet per second, its momentum is $\cdot 0621 \times 20$, or $1\cdot 242$. If this momentum is created or destroyed by a force acting for only one second, the force must be $1\cdot 242$ lb.; if it is created or destroyed by a force acting for 5 seconds, the force must be $1\cdot 242 \div 5$, or $\cdot 2485$ lb. The mass of a body multiplied by its velocity represents its momentum. Momentum is sometimes defined as the *quantity of motion* possessed by a body. It represents the constant force which, acting for one second, would stop the body. Suppose a certain amount of momentum is produced by a force of one pound acting on a body for one second, the same amount of momentum would be produced by a force of ·2 lbs. acting for half a second, or by 1,000 lbs. acting for the thousandth of a second, or by ·001 lb. acting for 1,000 seconds. You can easily prove this by what you already know about the motion of a body.*

If the velocity of a body has been produced or destroyed by various forces, each acting for a certain time, multiply each force by the time during which it acted and the sum must be equal to the whole momentum generated or destroyed. When we know the *time* during which a certain force has acted on a body giving to it motion, we generally determine the motion by calculating the momentum of the body. When we know the *distance* through which a force has acted on a body giving to it motion, we generally first find the kinetic energy of the body.

217. Impulse or Blow.—When a body is suddenly set in motion or stopped in its motion by a blow, the effect of the blow is measured by the total momentum produced or the total momentum destroyed. We know that great forces must have been acting for short times to produce the effect we observe, but it seems difficult to find the magnitude of each force and the time during which it acted. The average force during a blow may be approximately found by dividing the momentum produced or destroyed by the short interval of time during which the force acted.

218. Resultant.—The resultant of two or more forces is a force which might be substituted for them without changing the effect. If two strings pull a point with forces of 5 lbs. and 7 lbs. (Fig. 134), and if the angle between them is 30°, draw O P equal in length to 5 inches, and make the angle Q O P equal to 30°.

* If a force acts on a body, the acceleration which it produces is equal to the force divided by the mass of the body, and this acceleration multiplied by the time gives the velocity. Hence the force multiplied by the time equals the mass multiplied by the velocity which is produced in that time.

Make the length of O Q, 7 inches. Complete the parallelogram Q O P R, and draw the diagonal O R. Measure O R in inches; we find it to be 11·6 inches, so that the resultant of the two forces is 11·6 lbs. One string acting in the direction O R with a pull of 11·6 lbs. will produce the same effect at O as the two strings did.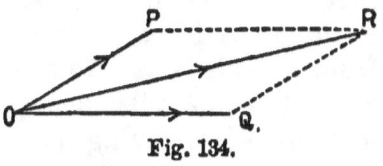

Fig. 134.

Suppose when the two strings were acting we had found by experiment that a third string O E (Fig. 135) would just prevent the two strings from causing motion at O, then experiment would also show that the force in O E, which may be called the *equilibrant* of O P and O Q, is exactly equal and opposite to the resultant of O P and O Q.

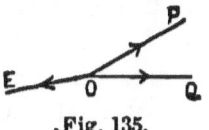

Fig. 135.

219. Equilibrium.—If three and only three forces keep a body in equilibrium, we know that these three forces must be parallel to and proportional to the sides of a certain triangle. If a number of forces keep a body in equilibrium, and if all their lines of action meet in a point, they must be parallel to and proportional to the sides of a certain closed polygon. But suppose they do not meet in a point, two important conditions hold: first, the forces are parallel to and proportional to the sides of a certain closed polygon, else they would make the body move as a whole; secondly, the turning moments of the forces about any axis whatsoever must balance, else the body would rotate. When the forces acting on a body are parallel to one another our rule becomes simpler, and we find, first, that the sum of all the forces in one direction must be equal to the sum of all the forces in the other direction; secondly, that the turning moments of the forces about any axis whatsoever must balance.

220. Moment of a Force.—This is the measure of the tendency of a force to turn a body about an axis. It is the force multiplied by the perpendicular distance between the direction of the force and the axis; in fact, the force multiplied by what has been called its leverage. A force of 5 lbs. acting at the distance of 2 feet from an axis is said to have a moment of 10 *pound-feet*,* and a force of 2 lbs. acting at the distance of 5 feet, or a force of 4 lbs. acting at the distance of 2·5 feet, would have the same turning moment.

If a number of forces tend to turn a body about an axis, find the sum of the moments tending to turn the body in the direction of the hands of a watch; find the sum of the moments tending to turn the body against the hands of a watch. The difference of these two sums is the moment which really acts on the body. If the two sums are equal the body will not turn.

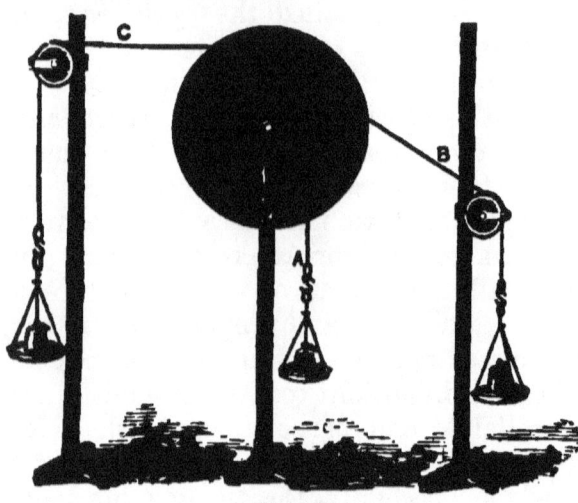

Fig. 136.

221. *Example.* — Fig. 136 shows a circular piece of wood which can turn easily about an axis. Suppose the cord AA, which is fastened by means of a nail to this wooden disc, is pulled by a weight of 4 lbs., and that when I measure accurately the perpendicular distance from the axis to the string I find it 1·2 foot. Then the moment of this force is 4 × 1·2, or 4·8 pound-feet in the direction of the hands of a watch. Suppose that I do the same for all the other cords which are keeping the wood at rest, I

---

* If the force does not act in a plane at right angles to the axis, it is only its resolved part in this plane which we consider.

can then make such a Table as the following, and find that the sum of the turning moments in one direction is equal to the sum of the moments in the other direction. This Table shows the result of a certain experiment in which six forces were acting.

Forces Tending to Turn Body in Opposite Direction to that of Hands of Watch.			Forces Tending to Turn Body in Same Direction as that of Hands of Watch.		
Force in Pounds.	Distance in Feet of Force from Axis.	Moment in Pound-feet.	Force in Pounds.	Distance in Feet of Force from Axis.	Moment in Pound-feet.
1·4	0·3	4·2	4·0	1·2	4·8
2·4	5·0	12·0	6·2	1·5	9·3
8·6	1·0	8·6	2·7	4·0	10·8
	Total	24·8		Total	24·9

If you try such an experiment as this, to complete it and to get a thorough understanding of the matter, give a *very small* turn to the wooden disc and measure the amount of rise or fall of every weight, using a magnifying glass and a very small scale. Some rise and others fall, but you will find that the *work given out* by those that fall is equal to the *work given to* those that rise. You will see that this is the reason why the above rule is true. Perhaps this will appear more evident if you imagine that, instead of being tied to separate nails, the cords had passed round pulleys of different diameters, but all fixed to one axle. Again, if you turn the wooden disc (Fig. 136) very slightly, you alter all the distances, so that there is no longer equilibrium. If you had pulleys this would not be the case; there would be equilibrium in any position.

222. **Torque.**—When a system of forces satisfies the first condition for equilibrium given above, but does not satisfy the second condition—that is, when it is equivalent to a couple—the system is called a *torque*. The

reason why this name has been introduced by Professor James Thomson is that such systems of forces have hitherto been vaguely called "a moment," "a couple," and so forth, and these names rather express the effect of the torque, or its resultant, than the torque itself.

223. A lever is a body with one point fixed, called the *fulcrum*. There is equilibrium when the turning moment of one of the balancing forces about the fulcrum is equal and opposite to the turning moment of the other. Thus a force of 1 lb. acting at 20 feet from the fulcrum will balance a force of 50 lbs. acting at 0·4 foot from the fulcrum.

224. If two parallel forces acting on a body are equal and opposite in the sense of their action, they form what is called a couple, and as they have no other tendency than to turn the body, they are always measured by their moment, for they evidently have the same moment about any axis we may choose. Thus, a couple consisting of two parallel forces, each of 8 lbs., if they are 3 feet asunder, is said to be a couple whose moment is 24 pound-feet. If they consisted of two forces, each of 12 lbs., 2 feet asunder, or of 6 lbs. 4 feet asunder, or of 24 lbs. 1 foot asunder, they would have exactly the same effect. Hence, when we speak of a couple we always speak of it as a couple of so many pound-feet.

225. Work.—To do work it is necessary to exert a force through a certain distance in the direction of the force. Thus, if I exert a force of 20 lbs. through a distance of 6 feet, I do 20×6, or 120 foot-pounds of work. If a body of 5 lbs. weight changes its level by the amount of 10 feet, whether it does this by a direct vertical fall or rise, or is moved up or down an inclined plane or curved surface, so long as there is no friction, the amount of work given out by the body in falling or given to it to make it rise is always the same, 5×10, or 50 foot-pounds.

226. *Example.*—One of the weights of a certain clock is 20 lbs., and after being wound up it can fall through a distance of 40 feet. Suppose we wish to alter this height, making it 10 feet; what weight must we use? Evidently the work given out by the new weight in falling 10 feet must be equal to the old weight 20 × 40 feet, or 800 foot-pounds. In fact, the new weight must be 80 lbs. Of course you must apply this weight to the clock by means of a block and pulleys, or you must reduce the diameter of the drum proportionately; and if in applying it you introduce more friction than there used to be in the clock, you must further increase the weight, so as to be able to overcome this friction.

227. Horse-power.—One horse-power is the work of 33,000 foot-pounds done in one minute. Remember that power really means, not merely work, but work done in a certain time. The work done in one minute by any agent divided by 33,000 is the horse-power of that agent. In a steam-engine, the piston travels four times the length of the crank in one revolution, and all this time it is being acted on by the pressure of steam. If the *mean* or average pressure urging the piston is 60 lbs. per square inch, and the area of the piston is 150 square inches, then the total average force urging the piston is 150 × 60, or 9,000 lbs. If the crank, whose length is 0·9 foot, makes 70 revolutions per minute, then the piston travels 4 times 0·9 × 70, or 252 feet per minute, so that the work done in one minute is 9,000 × 252, or 2,268,000 foot-pounds. Dividing this by 33,000, we find the horse-power of the steam-engine to be 68·7. The mean pressure is best found by the use of an *indicator* which draws for us an *indicator diagram*. Measuring the pressures at ten equidistant places on this diagram, adding them together, and dividing by ten, gives the average pressure. If an indicator diagram can be obtained from both sides of the piston so much the better, as we add our two results and divide by two. As the pressure of steam is usually given per square inch, it is usual to take the diameter of the cylinder in inches, but distances passed through by the piston are evidently to be measured in feet.

Exercise.—I find by a spring balance that some horses or a steam-engine have been pulling a carriage with an average pull of 120 lbs. during one minute, the space passed over in the minute being 500 feet; what is the horse-power expended on the carriage? Here 120 lbs. acts through the distance of 500 feet, and the work done in one minute is evidently 500×120, or 60,000; dividing by 33,000, we find the horse-power to be 1·818.

228. **Energy** is the capability of doing work. When a weight is able to fall, it possesses *potential* energy equal to the weight in lbs. multiplied by the change of level in feet through which it can fall. When a body is in motion, it possesses *kinetic* energy equal to its weight in lbs. divided by 64·4 and multiplied by the square of its velocity in feet per second.

229. *Example.*—A body of 60 lbs. is 100 feet above the ground, and has a velocity of 150 feet per second, What is its total amount of mechanical energy?—that is, what energy can it give out before it reaches the ground, and becomes motionless? Here the potential energy is 60×100, or 6,000 foot-pounds. Its kinetic energy is $60 \times 150 \times 150 \div 64\cdot 4$, or 20,963 foot-pounds. So that the total amount is 26,963 foot-pounds.

Exercise.—Suppose this body to lose no energy through friction with the air, and suppose that, after a time, it is at the height of 20 feet above the ground; find its velocity. Answer. Its *potential* energy is now 60×20, or 1,200 foot-pounds, therefore its *kinetic* energy must be 25,763, and evidently this, multiplied by 64·4 and divided by 60, gives the square of the new velocity, which I find to be 168·1 feet per second. Evidently in such a question we are not concerned with the direction in which the body is moving. It may be a cannon-ball, or a falling or rising stone, or the bob of a pendulum. Given its velocity and height at any instant, we can find for any other height what its velocity must be, or for any other velocity what its height must be.

230. Angle.—An angle is the amount of opening between two straight lines. An angle can be drawn: First, if we know its magnitude in *degrees*. A right angle has 90 degrees. Second, if we know its magnitude in *radians*. A right angle contains 1·5708 radian. Two right angles contain 3·1416 radians. One radian is equal to 57·2958 degrees. One radian has an arc, B C, equal in length to the radius A B or A C. It sometimes gets the clumsy name "a unit of circular measure." Third, we can draw an angle if we know either its *sine, cosine,* or *tangent,* &c. Draw any angle, B A C (Fig. 137). Take any point, P in A B, and draw P Q at right angles to A C. Then measure P Q and A P in inches and decimals of an inch. P Q ÷ A P is called the sine of the angle. A Q ÷ A P is called the cosine of the angle. P Q ÷ A Q is called the tangent of the angle. Calculate each of these for any angle you may draw, and measure, with your protractor, the number of degrees in the angle. You will find from a book of mathematical tables whether your three answers are exactly the sine, cosine, and tangent to the angle. This exercise will impress on your memory the meaning of these three terms.

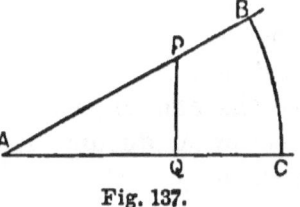

Fig. 137.

Divide the number of degrees in an angle by 57·2958, and you find the number of radians. Suppose we know the number of radians in the angle B A C, and we know the radius A B or A C, then the arc B C is

A B × number of radians in the angle.

Given, then, a radius to find the arc, or given an arc to find the radius, are very easy problems.

231. Angular Velocity.—If a wheel makes 90 turns per minute, this means that it makes 1·5 turn per second. But in making one *round* any radial line moves through the angle of 360 degrees, which is 6·2832 radians; so that 1·5 round per second means 6·2832 × 1·5, or

9·4248 radians per second. This is the common scientific way in which the *angular velocity* of a wheel is measured; so many radians per second. If a wheel makes 30 rounds per minute, its angular velocity is 3·1416 radians per second. One *round* is the angular space traversed in one revolution.

The real velocity in feet per second of a point in a wheel is equal to the angular velocity of the wheel multiplied by the distance in feet of the point from the axis.

232. Angular Acceleration.—The acceleration to angular velocity per second. If a wheel starts from rest, and has an angular velocity of 1 radian per second at the end of the first second, its average angular acceleration during this time is 1 radian per second per second.

233. *Comparison of Linear Motion and Angular Motion.*

The mass of a body is its weight divided by 32·2.	The moment of inertia of a wheel, or other rotating body is found by taking the mass of each portion of it and multiplying by the square of its distance from the axis.
A linear motion is given to a body when an unbalanced force acts upon it.	To produce angular motion—that is, rotation—it is necessary to have an unbalanced force acting at a distance from the axis of rotation. Force multiplied by perpendicular distance from axis is called the turning moment of the force.
Acceleration of a body is equal to force ÷ mass.	Angular acceleration of a body moving about an axis is moment of force ÷ moment of inertia.
If a force continually acts on a body, the velocity is equal	If a turning moment continually acts on a body (as by

Chap. XXI.]   LINEAR AND ANGULAR MOTION.   237

to acceleration multiplied by time from rest. Also

Space passed over is equal to half acceleration multiplied by square of time.

Energy stored up in a body is half its mass multiplied by square of its velocity.

If a body moves backwards and forwards under the action of a variable force which is always proportional to the distance of the body from its middle position, and which always acts towards this position, and if the force at a distance of one foot is 5 lbs., then the time of vibration is equal to 3·1416 times the square root of the quotient of the mass of the body divided by 5.

If a force of 20 lbs. acts on a body through the distance of 3 feet in the direction of the force, the work in foot-pounds done by the force on the body is equal to the force 20 multiplied by 3, or 60 foot-pounds.

If, then, a body receives power, say like a carriage, by a force acting on it in the direction of motion, the horse-power received is equal to the force in pounds multiplied by the distance in feet passed

a cord wound very many times round the axle of a wheel with a weight at its end), the angular velocity is equal to the angular acceleration multiplied by the time from rest. Also

The angle described by the wheel in any time is equal to half the angular acceleration multiplied by square of time.

Energy stored up in a wheel is half its moment of inertia multiplied by square of angular velocity.

If a wheel vibrates about its axis under the action of a spiral spring or twisted wire, so that the torque is always proportional to the angular displacement of the wheel from its mean position, and if the torque is 5 pound-feet when the wheel is 1 radian from the mean position, then the time of a vibration is equal to 3·1416 times the square root of the quotient of the moment of inertia of the body divided by 5.

If a torque is 30 pound-feet, and it turns a wheel through the angular distance of three radians, the work in foot-pounds done upon the wheel is equal to the torque 30 multiplied by 3, or 90 foot-pounds.

If, then, a body receives power, say through a shaft, the horse-power received is equal to the turning moment in pound-feet acting on the shaft multiplied by the angle in radians described in one minute (or the through in one minute divided by 33,000.

The mass of a body multiplied by its velocity is the *momentum* possessed by the body.

A force multiplied by the time during which it acts in hastening or stopping the motion of a body is equal to the momentum produced or destroyed.

If a body has momentum represented in direction and amount by the line O P (Fig. 138), and if a force acting in the direction O Q produces a change of momentum represented by the length of O Q, then O R is the resultant momentum in magnitude and direction possessed by the body after the operation of the force.

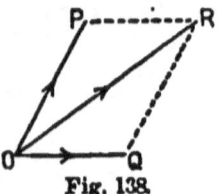

Fig. 138.

If forces represented in direction and magnitude by the lines O P and O Q act on a body, their action is the same as that of a force represented in direction and magnitude by the line O R.

number of rounds per minute × 6·2832), divided by 33,000.

The moment of inertia of a body multiplied by its angular velocity is its *moment* of *momentum*.

A torque multiplied by the time during which it acts in hastening or stopping the rotation of a body is equal to the moment of momentum produced or destroyed.

If a rotating body's axis is in the direction O P, and if its moment of momentum about this axis is represented by the length of O P, and if forces act upon it so as to turn it about an axis whose direction is O Q, and if the amount of moment of momentum produced by the torque is represented by the length of O Q, then the resultant motion of the body is a rotation about the axis O R, its moment of momentum being represented by the length of O R. (The arrow-heads in this case mean that an eye at O sees that the rotations are in the same sense—that is, all against the direction of the hands of a watch, let us suppose.)

If torques act on a body, if one of these is about an axis in the direction O P, and if the amount of the torque in pound-feet is represented by the length of the line O P, and if O Q similarly represents the torque about an axis in the direction O Q, their combined action is the same as that of a torque about an axis in the

| The change of momentum produced in a body which receives an impulse is equal to the sum of the products of the pressures during the impulse, each multiplied by the time during which it acts. | direction o R, the amount of the torque being represented by the length of the line o R. (The arrow-heads have the same meaning as in last case.) The change of moment of momentum produced by an impulse is equal to the sum of the products of the moments of the pressures during the impulse, each multiplied into the time during which it acts. |

234. Centrifugal Force.—If a body is compelled to move in a curved path, it exerts a force directed outwards from the centre, and its amount in pounds is found by multiplying the mass of the body by the square of the velocity in feet per second, and dividing by the radius of the curved path. Thus a weight placed at the end of an arm like the arm of a wheel exerts a pull in the arm. If a body moves round an axis 20 times per minute in a circle whose radius is 3 feet, you can determine the centrifugal force by first finding the velocity of the body and using the above rule, or you may proceed as follows:—The weight of the body multiplied by 3 multiplied by the square of 20 divided by 2,937 is the centrifugal force.*

Suppose a wheel, whose total weight is 20 tons or 44,800 lbs, has its centre of gravity 0·4 foot away from the axis—that is, suppose the wheel to be somewhat eccentric—then if the wheel makes 50 revolutions per minute, the centrifugal force is $44{,}800 \times 0{\cdot}4 \times 2{,}500 \div 2{,}937$, or 15,253 lbs.—that is 6·81 tons. This force acts on the bearings of the shaft, always in the direction of the centre of gravity of the wheel (see Chap. XX.).

235. Any one who wants to get clear ideas about

---

* Centrifugal force $= \dfrac{mv^2}{r}$, or ma^2r, or $wrn^2 \div 2{,}937$; where m is mass of body, or w its weight in pounds; r, radius of curved path; v, velocity in feet per second; a, angular velocity in radians per second; and n, number of revolutions per minute.

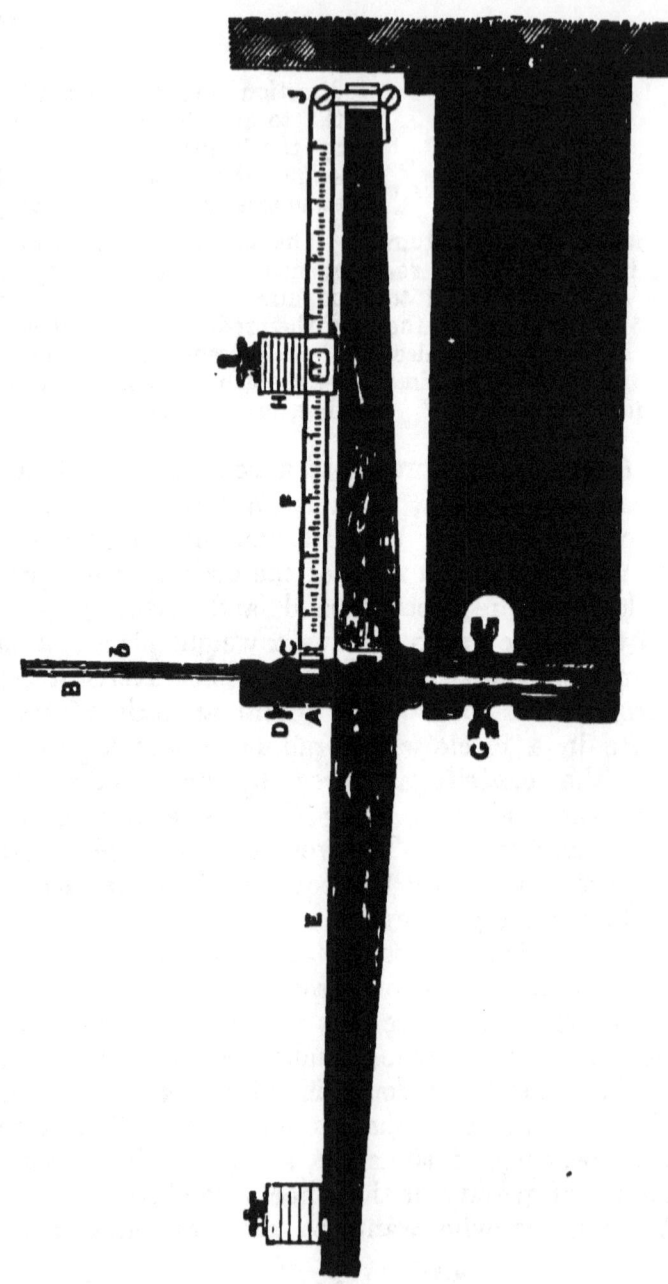

Fig. 139.

Chap. XXI.] EXPERIMENTS ON CENTRIFUGAL FORCE. 241

centrifugal force ought to make experiments of his own. Unfortunately, although there are many toys made to illustrate the effects of centrifugal force, I know of only one piece of apparatus which enables the laws to be systematically experimented upon. Fig. 139 is a drawing of such an instrument. Here A represents a little, flat, cast-iron box, like a narrow drum; one drum-head, as it were, being replaced by a corrugated steel plate, so as to be strong and flexible. B is a glass tube which enters the box. Mercury fills the box and the tube to the level b, and when C, the centre of the corrugated plate, is pulled or pushed, although you cannot see much yielding in C, you will observe the mercury rise and fall in the tube. There is a screw, D, entering the box at the back; by means of this screw you can adjust the height of the mercury in the tube. The box is in the centre of a circular table, E, which can be whirled round at any speed we please, and the tube is exactly in the axis of rotation, so that the height of the mercury can be measured whatever be the speed. Fastened to the centre of the corrugated plate at C is a long brass rod, F, which is supported at J on the end of a little rocker, so that it can move backward and forward with less friction than if it were made to slide on a bearing. At any place along F we can clamp a weight, H, which we may alter as we please from 0·5 to 8 lbs. We can clamp it near the axis or one foot away, the radius of the circle described by its centre of gravity being measured by the scale marked on the rod. We have also a means of counting the number of revolutions made per minute. Now, the centrifugal force due to the rod and sliding weight causes C to be pulled out very slightly, and this causes the mercury to fall in the tube, and it is easy with a vertical scale firmly attached to a neighbouring wall, but placed alongside the tube, to measure this rise and fall. I usually get a spring balance, or a cord, pulley and weights, and before my experiments begin I pull the end of the rod, F, noting the height of the mercury for a pull of 1 lb., 2 lbs., &c., and in this way I can afterwards tell the value of my scale measurements. I also make a number of experiments when the sliding weight is removed from the rod, F, to tell me the centrifugal force of everything else at different speeds, and this I subtract from my subsequent observations. You see, then, that I can measure the centrifugal force in pounds of various masses, from 0·5 to 8 lbs., moving at any speed in a circle whose radius may be as little as 1·5 inch and as much as 11·5 inches. With this instrument it is easy to test the law which is usually given, and without working with some such instrument I question if you are likely to have any but vague notions about centrifugal force.

236. There is another method of experimenting which suggests itself, with apparatus which any one may fix up for himself, but it does not give such a thorough understanding of the law to the person who experiments. In Fig 140, A is a leaden ball at the end of a silk thread, P A, fastened at P. A is kept out from its natural position in the vertical by means of a horizontal thread in the direction A B. Now if you pass the horizontal thread, A B, over a pulley and hang a weight at its end, you will find that the force acting in A B is to the weight of A as the distance K A is to the distance K P.* Now suppose the weight of A to be 4 lbs., the height, P K, to be 8 feet, and the distance, A K, 1 foot, then

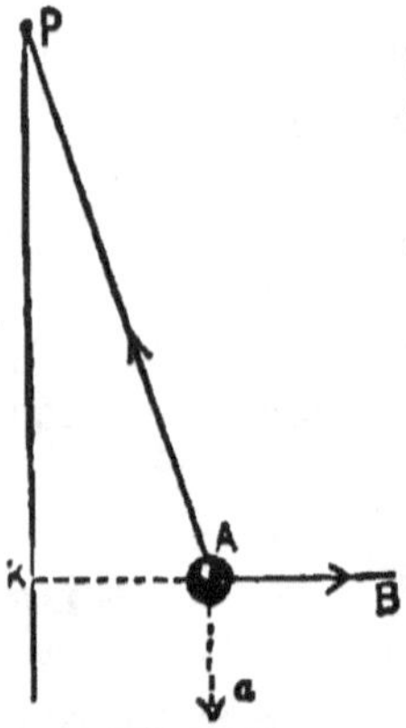

Fig. 140.

---

* The body, A, is acted upon by three forces: its weight downwards in the direction A a, the horizontal force, A B, and the pull in the string, A P. The triangle of forces (Art. 2) tells us that as A K P is a triangle whose sides A K and K P and P A are parallel to the three forces, then the horizontal force, acting in A B, is to the vertical force which is the weight of A, as the distance K A is to the distance P K.

A K is one-eighth of K P, and we are sure that the horizontal force needed just to keep A in this position is 0·5 lb., for it must be one-eighth of the weight of A.

Now let such a ball as A, hung by a thread, P A, go round and round in a circle. Measure as accurately as you can K A, the radius of the circle, and K P, the *vertical* height from the ball to the point of suspension. Also measure how many revolutions the ball makes per minute. The centrifugal force is now doing what the horizontal string did before, and we know how much it is. In fact, the centrifugal force is obtained by multiplying the weight of the ball by K A, the radius of its circle, and dividing by the vertical height, K P. You can test if the centrifugal force law is true, therefore, by means of your measurements.*

237. Friction.—When one surface slides on another, there is a force resisting the motion. This force is found

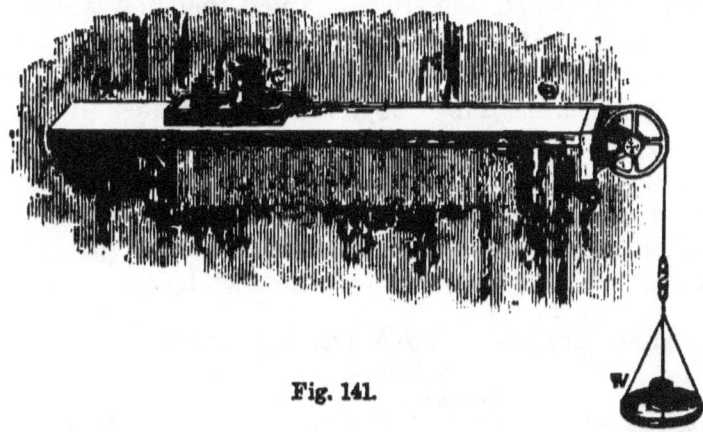

Fig. 141.

to be equal to the force pressing the surfaces together multiplied by a number which must be found experimentally for different kinds of surfaces. It may be

* A ball going round in the way above described is called a *conical pendulum*. You will find from observation that it makes one revolution in the same time as that in which an ordinary pendulum performs a complete oscillation backwards and forwards, if the length of the ordinary pendulum is equal to the vertical height from A to P or P K.

found experimentally as follows:—In Fig. 141, A B represents a table, the upper level surface of which is wood, iron, brass, or other material to be experimented upon. We usually experiment on smooth surfaces. C is a little slide made of any material whose coefficient of friction we wish to find. Different weights may be placed on it. The weight of the slide, together with the weight lying upon it, is the total force pressing the two surfaces together. C is pulled by the weight, W, hung from a string, passing over a pulley working on very frictionless pivots. The weight, W, which will just cause the slide to keep up a slow, steady motion on the table, is taken as a measure of the friction. Of course, however, it really includes the friction of the pulley, but this is usually neglected, as it is small. It is found necessary to start the slide by giving a little jerk to the arrangement, as, when one of the surfaces is of wood, the friction when the slide is motionless is somewhat greater than when it is moving. This is one of the most instructive experiments which can be made in mechanics, and I hope that every reader will make a series of observations. Let him correct his results by means of squared paper, and he will find that the friction is a constant fraction of the force pressing the surfaces together. This fraction is called the "*coefficient of friction.*" I give its value for a few surfaces.

Oak on oak, fibres parallel to direction of motion . . 0·48
„ „ perpendicular „ „ . . 0·34
„ „ endwise „ „ . . 0·19
Metals on oak „ parallel „ „ 0·5 to 0·6
Wrought iron on wrought iron, wrought iron on
  cast iron . . . . 0·18
Cast iron on cast iron . . . . 0·15
Smoothest and best greased surfaces on one another 0·03 to 0·036

It is very interesting, after determining the coefficient in the case of a certain pair of materials, to diminish the size of the slide. You will find that unless you diminish it so much that the pressure actually alters the surfaces in contact from being quite plane you will get pretty much the same result. You will also find that, whether the motion of the slide is quick or slow, if your weight maintains the motion steady when it is slow it will also maintain it steady when quick.*

238. You will find it instructive to experiment with such a piece of apparatus as is represented in Fig. 142, and which I designed in Japan to measure the friction between sliders of different materials and this cast iron wheel. Here we have a pulley with a broad, smooth outer surface. On this surface lies a slide made slightly concave, to fit the rim

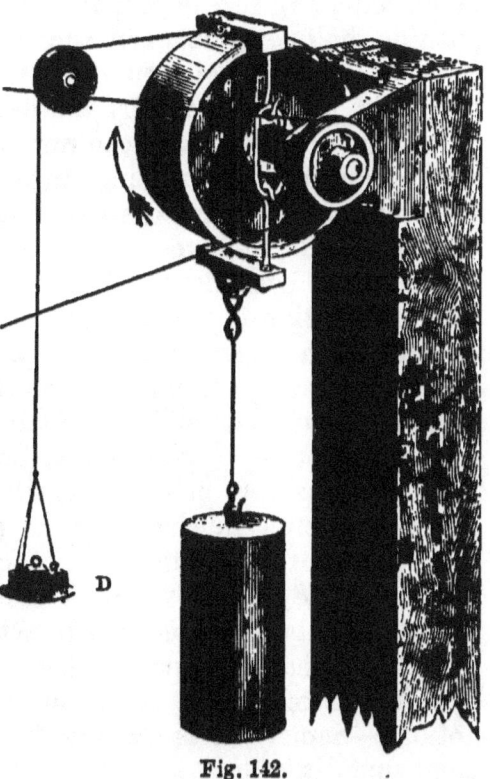

Fig. 142.

of the pulley. On this slide we can hang different loads by the arrangement shown in the figure, and the slide can only move a small distance in any direction on account of four guards. Suppose now that

---

* If, instead of using a cord and weight, we move the slide by tilting the table more and more from the horizontal, the slide getting an occasional shove to start it, let the inclination of the table be found in degrees when the weight of the body itself is just able to keep up a steady motion. The tangent of the angle of inclination of the table when this occurs can be found in a book of mathematical tables; it proves to be equal to the coefficient of friction. This method of experimenting is much easier and is more exact than the other, but it is not so instructive.

the pulley rotates in the direction of the arrow. Friction causes the slide to move in the direction of the arrow until it is brought up by the guard. Now let weights be placed in the scale-pan, D, until the slide is held in a position half way between the guards. Evidently the force of friction between the slide and the rim of the pulley is just balanced by the weight in the scale-pan. With this apparatus you can not only find the coefficient of friction for two rubbing surfaces easily, but you can very quickly vary your experiments. Run the pulley at various speeds, and you will always find much the same value for the coefficient of friction. Again you will find that the friction is proportional to the load on the slider.

But with an apparatus of this form you will be able to get very useful information of another kind. I have shown the slide as fitting the surface of the rim of the pulley. Now let it be flat, or even convex. We have reason to believe that when the rubbing surface becomes altered in shape by the pressure applied to it, the coefficient of friction is not quite the same for all pressures. By using slides of different forms, you will be able to submit this notion to actual experiment. I do not think that any person has carried out the idea, and, in this respect, it is like a great many other vague notions—namely, it is waiting for my readers to experiment upon it.

239. As it has been found that with some kinds of material the *statical friction*—that is, the friction which resists motion from rest—*is somewhat greater than the friction of the surfaces when actually moving*, experiments have been made to determine whether, at very small velocities indeed, with such materials, there is not a gradual increase in the friction. It is known that at ordinary velocities the friction is much the same as at a velocity of ·01 foot per second. We have reason to believe that with *metals on metals* there is the same friction at all velocities, even down to one-five-thousandth of a foot per second, whereas with *metals on wood* the

friction increases gradually as the velocity diminishes, until when the velocity is 0, the friction is what we call static friction. Again, at very high velocities it has been found that there is a very decided diminution of the coefficient of friction between a cast iron *railway brake* and the wrought iron tyre of a wheel. The coefficient was ·33 for very slow motion, ·19 for a speed of 29 feet per second, and ·127 for a speed of 66 feet per second. It has also been observed in these railway brake experiments that when a certain pressure is applied for a short space of time the friction diminishes. All such results as these, however interesting they may be to the railway engineer, tell us nothing about what I have hitherto called friction, because I have supposed the rubbing surfaces to remain unaltered, whereas these railway brakes are rapidly worn away, and the *effects of abrasion* and polishing are of an utterly different kind from the effects of friction of which I have hitherto been speaking.

240. You must remember that although friction leads to waste of energy, all the energy spent in overcoming friction being converted into another form of energy called *heat*, still **friction is often very useful.** The weight resting on the driving wheels of a locomotive engine multiplied by the coefficient of friction between the wheels and rails represents the greatest pull which the engine can exert upon a train. Suppose the weight on the driving wheels to be 15 tons; the coefficient of friction of wrought iron on wrought iron is about 0·2, hence the greatest pull which the locomotive can exert is $15 \times 0\cdot 2$, or 3 tons. If the train resists with a greater force than this, the driving wheels must slip; if the train resists with a less force than this, there is no slipping, the wheels simply roll on the rails. Again, it is the friction between the soles of our feet and the ground that enables us to walk; friction enables us to handle objects; friction enables a nail to remain in wood; friction keeps mountains from rolling down.

241. Fluid Friction.—I have been considering the

friction only between solid bodies, or solid bodies separated by very thin layers of liquid substances, such as oil or water. But the friction between liquids and solids or between liquids and liquids is of a very different kind. If a man attempts to dive into water unskilfully, and falls prone, you know that the water offers a very considerable resistance to a change of shape. (See Art. 193.) Now this is mainly the resistance that any body offers to being rapidly set in motion. If you came colliding against the end of the most frictionless carriage, you would also experience its resistance to being suddenly set in motion; whereas the constant steady resistance to motion which the carriage experiences when moving with a uniform velocity is called friction. What I wish rather to refer to is the resistance to the motion of water in a pipe, the resistance to the steady motion of a ship.

242. Fig. 143 shows a hollow cylindric body, F, supported so that it cannot move sidewise, and yet so that its only resistance to turning is due to the twist it would give the suspension wire, A. C C is water or other liquid filling the annular space between the cylindric surfaces D D and E E, and wetting both sides of F. When the vessel D D E E is rotated, the water moving past the surfaces of F tends to make F turn round, and this frictional torque is resisted by the twist which is given to the wire. The amount of twist in the wire gives us,

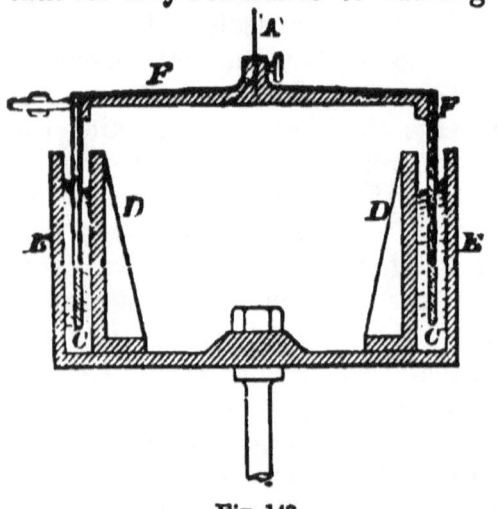

Fig. 143.

then, a measurement of the viscosity of liquids, and investigations may be made under very different conditions.

243. Fluid friction may also be measured by similar and equal heavy discs of brass, which you can immerse in air, water, and oil, and hang by wires from the ceiling. You will observe that, when the suspension wires are twisted and let go, the bodies vibrate like the balance of a watch. But it is only the one which vibrates in air that goes on vibrating for a long time; the one in water keeps up its motion longer, however, than the one in oil, showing that there is more frictional resistance in oil than in water, and more in water than in air. The *rate of diminution of swing* or the *stilling of the vibrations* tells us the viscous friction of the fluid. In fact, if, by means of a pointer or mirror attached to the wire, you observe the various angular displacements, noting the time for each, and then plot your observations on squared paper (as in Art. 187), you will find a *curve of sines* for the vibrations in air; and for the different liquids *damping curves*, which show the effect of friction in the liquids. Similarly, the rate of diminution of swing of the vibrating fluids in U tubes, one containing water and the other oil, tells us about the relative co-efficients of viscosity of the liquids. (See Exercise, page 205.)

244. From experiment, it is found that the force of friction in water is proportional to the wetted surface where friction occurs; it is proportional to the speed when the speed is small, but it increases much more quickly than the speed does. Thus, at the velocities of 1, 2, 3, &c., inches per second the friction is proportional to the numbers 1, 2, 3, &c., whereas at the velocities of 1, 2, 3 yards per second, the friction is proportional to the numbers 1, 4, 9, &c. At small velocities, such as the above mentioned cylinders experience, three times the speed means three times the friction; whereas, at great velocities, such as those of ships, three times the speed means nine or more times the friction. You see, then, that friction

in fluids is proportional to the speed when the speed is small, to the square of the speed when the speed is greater, and at still greater speeds the friction increases more rapidly than the square of the speed. Even in such a fluid as air, the resistance to motion of a rifle bullet is proportional to the cube of the speed. That is, a bullet going at twice the velocity, meets with eight times the frictional resistance from the atmosphere. Again, it has been found that the friction is much the same whatever be the pressure. Thus it is found that when the disc and liquid apparatus is placed in a partial vacuum or under considerable pressure there is exactly the same stilling of the vibrations.

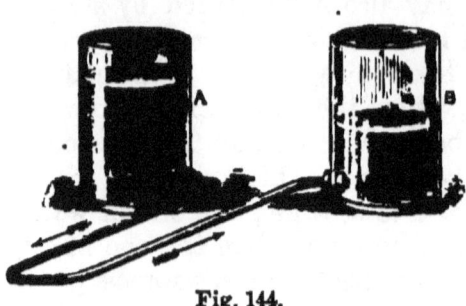

Fig. 144.

This fact is illustrated by the apparatus, Fig 144. Water tends to flow from vessel A to vessel B, through the long tube. Whether the tube is in the position shown in Fig. 144, or in the position Fig. 145, or is acting as a syphon, we find the same flow through it; the same quantity of water passes through it per second, although the pressure of the water in the tube in the position Fig. 145 is very much greater than in the position Fig. 144, or again when the tube is a syphon. The comparison is most readily made by observing how long it takes for a certain change of levels

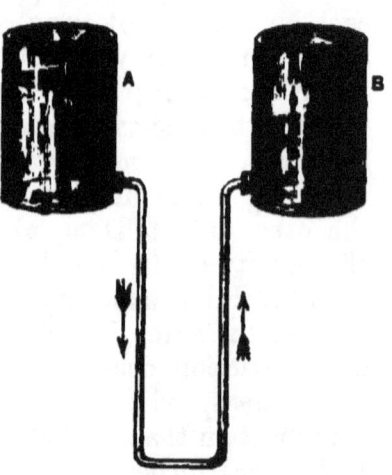

Fig. 145.

to take place in the two vessels, repeating this several times with the tube in various positions, beginning and ending each experiment with the same difference of levels. Again, fluid friction, for even considerable velocities, does not seem to depend much on the roughness of the solid boundary. This seems to be due to the fact that a layer of fluid adheres to the solid surface and moves with it. Even when the disc of Art. 243 is indented, or when large grooves are cut in it, we find practically the same frictional resistance.

245. *Comparison of the Laws of Fluid and Solid Friction.*

Friction between Solids.	Fluid Friction.
1. The force of friction does not much depend on the velocity, but is certainly greatest at slow speeds.	1. The force of friction very much depends on the velocity, and is indefinitely small when the speed is very slow.
2. The force of friction is proportional to the total pressure between two surfaces.	2. The force of friction does not depend on the pressure.
3. The force of friction is independent of the areas of the rubbing surfaces.	3. The force of friction is proportional to the area of the wetted surface.
4. The force of friction depends very much on the nature of the rubbing surfaces, their roughness, &c.	4. The force of friction at moderate speeds does not depend on the nature of the wetted surfaces.

APPENDIX.

RULES IN MENSURATION.

An area is found in square inches if all the dimensions are given in inches. It is found in square feet if all the dimensions are given in feet.

Area of a parallelogram.—Multiply the length of one side by the perpendicular distance from the opposite side.

The *centre* of *gravity* of a parallelogram is at the point of intersection of its diagonals.

Draw a *right-angled triangle;* measure very accurately the lengths of the sides. You will find that, no matter what scale of measurement you use, the square of the length of the hypotenuse is equal to the sum of the squares of the lengths of the other two sides.

Area of a triangle.—Any side multiplied by its perpendicular distance from the opposite angle and divided by two.

The *centre* of *gravity* of a triangle is found by joining (Fig. 146) any angle, A, with the middle point, D, of the opposite side, B C, and making D G one-third of D A. G is the centre of gravity.

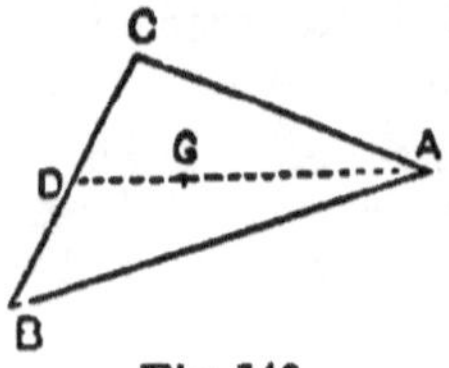

Area of any irregular figure.—Divide into triangles and add the areas of the triangles together.

Fig. 146.

Circumference of a circle.—Multiply the diameter by 3·1416.

Arc of a circle.—From eight times the chord of half the arc subtract the chord of the whole arc, one-third of the remainder will give the length of the arc, nearly.

Area of *circle.*—Square the radius and multiply by 3·1416.

Area of a *sector* of a circle.—Multiply half the length of the arc by the radius of the circle.

Area of a *segment* of a circle.—Find the area of the sector having the same arc, and the area of the triangle formed by the chord of the segment and the two radii of the sector. Take the sum or difference of these areas as the segment is greater or less than a semicircle.

Otherwise, for an approximate answer:—Divide the cube of the height of the segment by twice the chord, and add the quotient to two-thirds of the product of the chord and height of the segment. When the segment is greater than a semicircle, subtract the area of the remaining segment from the area of the circle.

The *areas of Indicator Diagrams and such curves* may be found by Simpson's rule.—Divide the area into any even number of parts by an odd number of equidistant parallel lines or ordinates, the first and last touching the bounding curve. Take the sum of the extreme ordinates (in many cases each of the extreme ordinates is of no length), four times the sum of the even ordinates, and twice the sum of the odd ordinates (omitting the first and last); multiply the total sum by one-third of the distance between any two successive ordinates.

Surface of a sphere.—Multiply the diameter by the circumference.

Surface of a *cylinder.*—Multiply the circumference by the length, and add the area of the two ends.

Surface of *any right pyramidal or conical body.*—Multiply half the circumference of the base by the slant side, and add the area of the base.

Lateral surface of the *frustum of a right cone.*—Multiply the slant side by the circumference of the section equidistant from its parallel faces.

Area of an *ellipse.*—Multiply the product of the major and minor axes by ·7854.

The *cubic content* of a body is calculated in cubic inches if all the dimensions are given in inches; in cubic feet if all the dimensions are given in feet.

Cubic content of a *plate.*—Multiply area of plate by its thickness.

Cubic content of a *sphere.*—Cube the diameter and multiply by ·5236.

Cubic content of the *segment of a sphere.*—Subtract twice the height of the segment from three times the diameter of the sphere; multiply the remainder by the square of the height, and this product by ·5236.

The cubic content and surface of a sphere are each two-thirds of that of the cylindric vessel which just encloses it.

Cubic content of *any prismatic body* (Fig. 147).— Multiply the area of the base by the perpendicular height. This will give the same product as, Multiply the area of cross section by length along the axis of the prism (the axis of a prismatic body goes from centre of gravity of

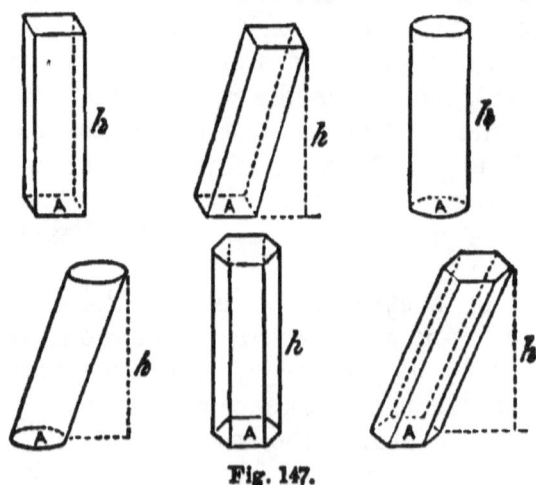

Fig. 147.

base to centre of gravity of top). The centre of gravity of a prismatic body is half way along the axis.

Cubic content of *any pyramidal or conical body*

(Fig. 148).—Multiply the area of the base by one-third of the perpendicular height.

Centre of gravity is one-quarter of the way along the

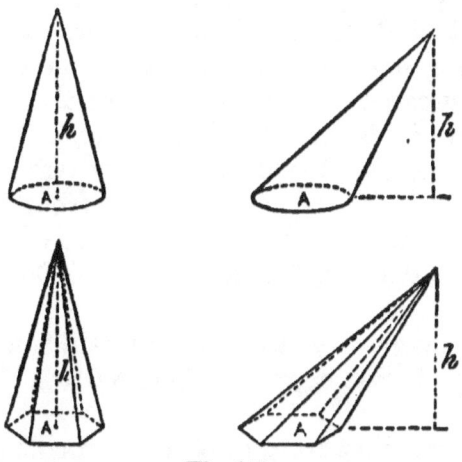

Fig. 148

axis from the base. (The axis of any such body joins the centre of gravity of base with the vertex.)

The cubic content of the *rim of a wheel* is found by multiplying the area of a cross section by the circumference of the circle which passes through the centre of gravity of the cross section.

The weight of a cubic inch of cast iron is about 0·262 lb. The weight of a cubic inch of wrought iron is about 0·28 lb. The weight of a cubic inch of brass is about 0·303 lb.

Hence, to find the weight of a cast iron body or a wrought iron body, or a brass body, first find the content in cubic inches and multiply by one of the above numbers, and we get the weight of the body in pounds.

Very often it is only the approximate weight that is wanted; so that a moulder may know how much metal to melt, or for other purposes.

Now suppose we want the approximate weight of a cast iron beam. Find roughly the average section, and get its area in square inches, multiply by the length in inches, add to this the cubic content of any little gusset plates or other excrescences. Multiply by ·262, and we have the weight in pounds.

You will find tables of specific gravities in many books. The *specific gravity* of a substance means its weight as compared with the weight of the same bulk of water. Now, it is known that a cubic foot of water weighs very nearly 1,000 ounces, or rather 62·4 lbs. The specific gravity of brick varies from 2 to 2·167, and therefore the weight of a cubic foot of brick varies from 2 × 1,000 or 2,000 to 2·167 × 1,000, or 2,167 ounces.

The specific gravity of cast copper is 8·6, and therefore a cubic foot of cast copper weighs 8·6 × 1,000, or 8,600 ounces.

We see, then, that from a table of specific gravities we can get the weight of a cubic foot of a substance, and therefore if we know the cubic content of a body formed of this substance we can calculate its weight.

Various *plans for saving labour in calculation* suggest themselves to people working at any particular trade. For instance, if a pattern has no prints for cores, the weight of the pattern bears nearly the same proportion to the weight of the casting as the weight of a cubic inch of the wood bears to the weight of a cubic inch of cast iron. This is not quite true, because the pattern is a little larger than the casting.

The *area of an irregular figure* may be obtained approximately by cutting it out of a uniform sheet of cardboard and weighing it. Now cut out a rectangle or square whose area it is easy to calculate. Weigh this also. The areas are in the same proportion as the weights.

The *area of cross section of a fine wire* can be determined with some accuracy by weighing a considerable length of the wire, dividing by the weight of the material per cubic inch, and dividing by the length of the wire in inches.

INDEX.

ABSORPTION dynamometers, 37, 38.
Acceleration, Definition of, 225
— — — angular, 236
— in pure harmonic motion, 181
Advantage, Mechanical, 14
— — of various machines, 17—26
Algebra, Necessity for knowledge of, 2
Alloys of copper, 85
Alternate pull and push, Effect of, 65
Amplitude of a vibration, 201
Angle, Definition of, 235
— Cosine, sine and tangent of, 235
— of twist, 31, 93
— Re-entrant, a weak point, 82, 99
Angular acceleration, Definition of, 236
— motion and linear motion, compared, 236
— velocity, Definition of, 235
Annealing, 83, 84
Apparatus for centrifugal force experiments, 240
— — fluid friction experiments, 248
— — kinetic friction experiments, 245
Arbor of spiral spring, Forces acting on, 167
Arch fitted to withstand fluid pressure, 111

Arch, Fuller's method of drawing link polygon for, 164
— How load is distributed over, 162
— ring, Forces at section of, 120
Arched rib in unstable equilibrium, 161
Attwood's machine, Arrangement of, 41
Axes of oscillation and suspension interchangeable, 197
— Principal, of an ellipse, 144, 253
Axis, Neutral, 105
— — passes through the centre of gravity, 105
— of oscillation, 195
— of suspension, 195, 217
— permanent, Meaning of, 219
— — Necessity for, 219
Axle, Friction of, 13
— wheel and, Mechanical advantage of, 20

BALANCE of watch, how compensated, 194
— — — Motion of, 193
— — — Rule for periodic time of, 193
— weights on wheels of locomotive, 221
Balancing of machines, 218, 219

R

Balancing of locomotive, Rules for, 222
Ballistic pendulum, 217
Beam, Behaviour of when loaded, 101
— Deflection of, 116, 128, 145
— — — Table of values, 116
— — — Formulæ for, 122, 123
— fixed at ends, 113, 119
— flanges to resist bending moment, 54, 106, 115
— loaded in various ways, 116–119
— Methods of supporting, 113
— of uniform strength, 119
— Rule for breaking load on, 120
— Strength of similar, 134
Bearings of shafts, Friction at, 12
— — — Pressure on, due to centrifugal force, 218
Beech timber, 78
Behaviour of materials when overstrained, 63, 100, 193
Belt, Difference of pulls in, 32
— Horse-power transmitted by, 32
Bending, 100
— and twisting, Relation between, for cylindric shafts, 112
— moment, 103
— — at section of beam, 104, 106
— — Diagrams of, 144, 146
— — — How to draw, for loaded beam, 114
— — — Table with, 116, 118
— — in a spiral spring, 168
— — Iron flanges of beams to resist, 115
— — of india-rubber beam, 101
Bifilar suspension, 200
Blow, 228
— Effect of, 206
— Motion produced by, 211, 216
Bodies, falling, Laws of, 39, 40
Boilers, Rule for bursting pressure (f, 5)

Boilers, Strength of spherical, 60
Boiler-plates, Riveted, 60, 88
Brass, Composition and use of, 85
Breaking load for beams, how calculated, 120
— — — struts or columns, Rules for, 131, 133
Bricks, Manufacture of, 72
— Characteristics of good, 72
Bronze, Composition and use of, 85
Bulk, Modulus of elasticity of, K, 57, 58
Bullet, Velocity of, how measured, 218
Buttresses, 165

CANDLE and board experiment, 213
Carbon in cast iron, 80
— — steel, 84
Case hardening of iron, 84
Castings, 80
— Chilled, 83
— Malleable, 83
— The cooling of, 81
Cast Iron, 80—83
— — beam, Flanges in, 54, 115
— — Closeness of structure of, 66, 80
— — Effects of carbon in, 80
— — Factors of safety for, 121
— — Grey, 80
— — Modulus of elasticity of bulk, K, for, 57
— — Strains in, due to inequality in rate of cooling, 82
— — Strength of, 68, 69, 131
— — Struts of, 54, 130
— — Toughened, 83
— — White, 80
Cedar timber, 78
Cement, 72
Centre of gravity, 252
— — — graphically determined, 141

INDEX.

Centre of gravity of section of bent beam is in the neutral axis, 105
— — — of wheels should be in axis of rotation, 219
— — — Ordinary formula for, 142
Centre of percussion, 195
— — — how found, 217
Centres of oscillation and suspension, interchangeable, 197
Centrifugal force, Apparatus for experimenting on, 240, 242
— — Definition of, 239
— — Effects of, on bearings of shafts, 218, 239
— — Rule for, 220, 239
Chain, Loaded, How to draw the curve in which it hangs, 160, 161
Chemistry, Useful, 70
Circle, Pitch, of spur wheel, 27
Coefficient of friction, 35, 244
Columns, 53, 54
— breaking stress for, Table of, 131
— Gordon's rules for strength of, 131, 133
— Hollow cylindric, of cast iron, 54, 59
— Long, break by bending, 54, 130
— Mode of breaking varies with length, 130
Combination of pure harmonic motions, 189, 192
Comparison of laws of fluid and solid friction, 251
— — linear motion and angular motion, 236, 239
Compensation balance of watch, 194
— pendulum, 185
Compound harmonic motions, curves of, and how plotted, 190—191
— — —, how produced, 189, 190, 192
— pendulum, 194
— — Rule for time of vibration, 195

Compression of struts or columns, 53, 58, 130
— — loaded beam, 102, 129
Concrete, Composition of, 73
Conglomerates, 71
Conical pendulum, Method of experimenting with, 242
— — Motion of, 179, 193
Connecting-rod and crank, Motions of, 28—30
— — Centrifugal force of, 220
Constraint of spiral spring, 197
— — twisted wire, 199
Conversion of energy, continual, 40, 187
Cooling of castings, 81—83
Copper, Alloys of, 85
— Properties and use of, 85
Cord, friction between post and, 33—35
Core used in castings, 81
Correction of errors in experiments, 10
Couple, Definition of, 232
— equivalent to a system of forces, 140
Couplings, 31
— Dynamometer, 31, 36
Crane, Efficiency of, 16
— Friction in, 15
— Law for, 17
Crank and connecting-rod, 28, 220
— — travel of piston, 233
Curvature, 223.
— Change of, in a spiral spring, 167
— of bent beam, or strip, 108
Curve, Elastic, 109
— of loaded chain, 160
— — sines, how drawn, 189, 200
— — — interpreted, 200
— — — traced by pendulum bob, 201
Curves showing elastic limit of loaded beams, 55, 128

Cutting and chipping, 65, 207
Cycloidal teeth give uniform motion, 27, 28

DAMPED vibrations, Law for, 201
Damping of vibrations, Investigation of, 201
— — — Representation of, 204
Decimals, Necessity for student to know, 1
Deflection of beams, 122—128
— — — Curves showing, 127
— — — Examples of, 123, 125
— — — Formulæ for, 122, 123
— — — loaded and supported in various ways; table of relative values, 116—118
— — — method of measuring, 124
— — — of different materials, table, 121
— — — whose sections are not rectangular, 126
— — similar beams similarly loaded, 134
Diagram of bending moment, 144
— — — — Table of, 116—118
— — Link polygon, 164
— Examples of stress, 155
Differential pulley-block, Mechanical advantage of, 19
Diminution of swing, Rate of, a measure of viscosity, 199, 205, 249
Discharge of water from orifices and pipes, 76, 77
Discs, Experiments with, 205, 249
Distribution of load on an arch, 162
Drawing, Experience to be gained by, 28
— instruments, necessary for student, 1

Drawings, Skeleton, 28-30
Dynamometer coupling, 31
Dynamometers, Transmission and absorption, 35—38

E, YOUNG'S modulus of elasticity, 54, 69, 92
Earth, Pressure of, against a wall, 73, 74
Eccentric-rod, 29
Edge, Re-entrant, ought to be rounded, 82, 99
Effect of friction, 6
— — a blow, 206
Efficiency of machines, 16
Elastic curve, 109
— — how drawn, 110
— — Examples of, 111
Elastic strength, 55, 59, 62
Elasticity affected by state of strain, 62, 63, 99, 100
— Law of, 55, 128
— Limits of, 55, 100, 128
— Modulus of, of bulk, K, 57, 58, 92
— — — — Table of values, 57
Electricity, Some knowledge of, useful, 70
Elm timber, 78
Ends fixed, Strength of beams with, 113, 119, 120
— — Columns with, 133
Energy communicated by a blow, how calculated, 216
— Communication of, dependent on shape of body, 212
— Continual conversion of, 40, 187
— Indestructible, 6, 39, 234
— in pile-driver, 208
— — rotating body, 42—50, 237
— — — — how determined, 45
— — waterfall, 6
— Kinetic, 38—50, 75, 234, 237

INDEX. 261

Energy, Loss of, due to friction, 11—14
— Potential, 38—40, 75, 234
— Storage of, during impact, 212
— — Store of, in water, 75
— wasted in fluid friction, Rule for finding, 76
Engine, Locomotive, balancing of, 220
— Steam, fly-wheel for, 50
— — giving out energy, 26
— — Indicated horse-power of, how found, 233
Equilibrant of two or more forces, 137, 229
Equilibrium of forces, 3, 139, 159, 229
Equivalent simple pendulum, 194
Escapements, 188
Experience gained by making skeleton drawings, 28
Experiments, Necessity for, 3
Extension, 51—58
— of part of loaded beam, 102, 108
— of similar and similarly loaded rods, 134
— of spiral spring, experiments, 174
— — wrought iron tie-rod, example, 53
— produced by load suddenly applied, 63
— proportional to load which produces it, 52

FACTOR of safety, 120
— — — Table giving usual values of, 121
Falling bodies, Laws of, 39—40
Fatigue of materials, 63, 100, 193
Figures, Properties of straight-line, 146, 147
— Reciprocal, 148

Firwoods, 77
Fixed ends, Beams with, 113, 119
— — Struts with, 133
Flanges in beams and girders, 54, 106, 115
Flat plates, Strength of, 133
Flow of water in pipes and pumps, 75, 77
— — — from an orifice, 76
Fluid friction, 75, 199, 247—251
— — Apparatus for investigating, 248
— — Laws of, 251
— pressure, 59, 61, 74, 75, 111
— — Arches to withstand, 111
Fly-wheels, Calculation of the size of, 48—50
Foot-pound, 12, 232
Force, Centrifugal, 218, 239
— how represented, 136
— Moment of a, 23, 229
— polygon, 137, 139
— Shearing, in beams and girders, 104, 114, 129
Forces acting at a point, 136—140, 159, 228
— — on arbor of spiral spring, 167
— Equilibrium of, 3, 137, 139, 229
— Polygon of, 4, 137, 229
— Triangle of, 3, 229
Friction and abrasion, 247
— a passive force, 4—5, 6, 243
— apparatus, 245
— at bearings of shafts, 12—14
— between cord and post, 33—35
— Coefficient of, 35, 244
— Effect of, as distinguished from force of, 6, 11
— Experiments on, 7—10
— force of, 11, 13
— in fluids, 75, 199, 247, 251
— — — measured by heavy discs, 205, 249
— — machines, 5—17
— — parallel motion, 13

Friction in quick-moving shafts, 14
— Kinetic less than statical, 246
— Law of, 4, 10, 15
— Laws of fluid and solid, compared, 251
— Loss of energy due to, 11—17, 89, 247
— never negligible, 44
— of metals on metals, 246
— — — — wood, 247
— often useful, 247
— proportional to pressure, 11, 243
— wheels, 13

GATES used in casting, 81
Geology, Principles of useful, 71
Girders (see Beams)
Glass, Composition and properties of, 79
— Toughened, 80
Graphical statics, 135—166
— — Mr. Bow's notation in, 150
Gravity, Intensity of, how found, 195
Gun metal, 85
Gun, Recoil of, 209
Gyration, Radius of, 144
Gyrostat, 215

HARMONIC motion, Pure, 29, 179
— — — acceleration at any point, 181
— — — Examples of, 180
— — — of liquid in U tube, 186
— — — — spiral-spring weighted, 183
— — — periodic time, Rule for finding, 182
— — — Representation of, 180

Harmonic motion, Pure, velocity at any point, 180
— motions, How to combine, 189, 190
— — — plotted, 190
Heat, Elementary principles of useful, 70
Heavy disc vibrating in fluids, 205, 249
Hinged structures, 146, 149
— — Calculation of stresses in, 149
Horse-power, 16, 233
— — of steam-engine, how indicated, 233
— — transmitted by belt, 32
— — — — shafts, 26—30, 237
— — — — toothed wheels, 133
— — — through coupling, how measured, 31
Hydraulic press, 25, 26
Hydrostatic arch, 111

IMPACT, average force of, 203
— mean pressure during, 211
— mutual pressure during, 208
— of two free ivory balls, 211
— Total momentum unaltered by, 208
Impulse and blow, 206, 208, 239
Inclined plane, 17, 18, 20
India-rubber beam, Bending of, 101
Indicator diagram, 233
Inertia, Moment of, 46, 105, 236
— — for rectangular or circular section, 107
— — greatest and least for any area, 144
Internal strains due to contraction in cooling, 79—83
— — at sharp corners, 82
Iron, Annealing of, 83, 84
— Cast, 80—83

INDEX.

Iron, Wrought, Varieties and properties of, 83, 84
Isochronism of spiral springs, 170

JOINTS, Effect of stiffness of, 156
— masonry, Middle third of, 130, 163, 166
— of arch, Link polygon nearly normal to, 163
— — structures, Stresses at, 149
— riveted, Strength of, 83
— Treenails in wooden, 78

KILLING wire, Meaning of, 62
Kinetic energy, 38—50, 234, 237
— — converted into potential, 40, 187
— — stored up in any machine, 50
— — — in water, 75
— friction apparatus, 245

LADDER, Forces acting on a, graphically determined, 141
Larch timber, 77
Law for a machine, 17
— of moments, 23, 230
— — — applied to stresses at section of loaded structure, 158
— — work, 17, 23, 231
Laws connecting variable things, how found, 7, 10
— of falling bodies, 39, 40
— of friction between solids, 251
— — fluid friction, 251
Level surface, Definition of, 223
Lever, 23—26, 222
Limestones, Compact and granular, 71

Limestones, Pure and hydraulic, 72
Limits of elasticity, 55, 100, 128
Linear motion and angular, compared, 236—239
Lines, and what they may represent, 136
Link motion, 30
— polygon, 137
— — a diagram of bending moment, 164
— — for arch, Fuller's method of drawing, 164
— — for minimum thrust at crown of arch, 164
— — must cut joint of arch nearly normal, 163
— — pole of, 138
Liquid in a U tube, Harmonic motion of, 186
— — — — Time of vibration of, 187
Load, Breaking, Rule for finding, 120
— carried by an arch, how distributed, 162
— Live or dead, 113, 121
— proportional to strength modulus, 106, 126
— Similar, on similar structures, 134
Loaded beam, Bending moment in, 114
— chain of suspension bridge, 160, 161
— links, Stresses in, 159
Locomotive engine, Balancing of, 220—222
— — Considerations in designing, 220

M OF A FLY-WHEEL, 44--49
— of any machine, 50
— ratio of values for similar wheels, 49

M, Use of, in designing fly-wheels, 48
Machines, Balancing of, 218
— Efficiency of, 16
— in box, 5
— Law for, 17
— Mechanical advantage of, 17—26
— Steadiness of, 48
Magnet, Time of vibration of, 200
Mahogany timber, 78
Mainspring in time-keepers, Office of, 187
Marble, Formation and character of, 71
Masonry arch, 162
Mass, Definition of, 223, 236
— Reciprocating, in locomotive, 220
Materials, Behaviour of, when overstrained, 62, 63, 99, 193
— Disposition of, in beams or girders, for strength, 106, 115, 119
Mean pressure during impact, 211
Mechanical advantage, 14
— — of blocks and tackle, 17
— — — differential pulley-block, 19
— — — hydraulic press, 26
— — — inclined plane, 18, 20
— — — lever, 23
— — — screw, 18
— — — wheel and axle, 20
Mechanics, method of study, 2, 28, 71
Mechanism, 26
Memel timber, 77
Mensuration, Rules in, 252
Metals, 80—85
— Tables giving strength of, 68, 69, 121, 131
Middle third of joint in arch ring, 130, 163
Modulus of elasticity of bulk, K, 57, 58, 92
— — — Young's, E, 54, 69, 92

Modulus of rigidity, N, 69, 87, 90, 92
Moment, Bending, 103, 104
— — Diagrams of, 116—118
— of a force, 23, 229
— — inertia, 46, 105, 107, 236
— — — Formula for, 142
— — — Poinsot's theorem regarding, 144
— — momentum, defined, 238, 239
Moments, Law of, 23, 230
— Method of, applied to stresses in structures, 158
Momentum, Definition of, 227
— Total, unaltered by impact, 203
Mortar, how prepared and why it hardens, 72, 73
Motion, Communication of, in liquids, 211
Motion is either translation, rotation, or both, 216
— Linear and angular, compared, 236—239
— Periodic, 179, 188
— — Rotational, 192
— Precessional, Examples of, 215
— produced by a blow, 216
Moulds for castings, 81
Muntz metal, 85
Mutual pressure during impact, 208

NATURE of pure shear strain, 90, 91
— — strain, 56, 57
Neutral axis, defined, 105
—:— passes through centre of gravity of section, 105
— — Radius of curvature of, 108
— line, 102, 105, 107
— surface, 102
Notation in graphical statics, Mr. Bow's, 150

OAK timber, 78
Orifice, Flow through, 76
Oscillation, Centre of, 195
— and suspension, Centres of, 197

PAPER, Squared, 7, 10, 15
Parallel motion, Friction of, 13
Patterns for moulds, 81
Pendulum arranged to trace curve of sines, 201
— Ballistic, 217
— Blackburn's, for combining harmonic motions, 189, 190
— Compensation, 185
— Compound, 194
— Conical, 179, 242
— Equivalent simple, 194
— Radius of gyration of, 193
— Simple, 37, 184, 187, 194
— — Time of vibration of, 185
Percussion, Centre of, 195
— — — how found, 217, 218
Periodic motion, 179
— — Examples of, 188
— — Rotational, 192
— time, 179
— — Rule for finding, 183
— — of balance of watch, 193
Permanent axes in machines, Necessity for, 219
— axis, Meaning of, 219
— set, 55, 100
Phosphor bronze, Composition of, 85
Pile-driver, Energy in, 206
Pillars, Gordon's formulæ for strength of, 131, 133
— Ways of breaking, 130
— with ends fixed, 131
— — — rounded, 133
Pipes, Flow of water in, 75

Pipes, Rule for bursting pressure of, 60
— Strength of water, 59
Piston, 28, 233
Piston-rod slide, Friction of, 13, 14
— — Strength of, 65
Pitch circle of spur wheel, 27
— of screw, 18
— — teeth, Rule for, 133
— — wheel teeth, 27
Plane, Inclined, 17, 20
Plates, Boiler, 60, 88
— Strength of flat, 133
Pole of link polygon, 138
Polygon, Closed, 137
— Force, 137
— Link, 137
— Unclosed, 137
— of forces, 4, 229
Potential energy, 38, 234
— — stored up in water, 75
— — converted into kinetic, 40, 187
Power, meaning of the term, 15, 233 (see Horse-power)
Precessional motion, 215
Pressure during impact, 208
Pressure-energy of water, 75
— Fluid, 59, 61, 75, 111
— on bearings of shafts, 218
— on teeth of wheels, 133
— of steam on piston, how found, 233
— — wind on roofs, 154
Principal axes of an ellipse, 144
Prints on patterns, 81
Propeller screw, Exercise on, 19
Puddling, 83
Pull in belt, 32
— how transmitted by wire, 51
— in cord, Difference of, on two sides of pulley, 4, 11, 32
Pulley and cord, Friction between, 33

Pulley block, Differential, mechanical advantage of, 19
Pumps, Flow of water through, 75, 76
Punching and shearing, 88
Pure harmonic motion of piston-rod, 29
— — — defined, 179 (see Harmonic motion)
— shear strain, 89
— — — how produced, 89
— — — Nature of, 90

QUASI-RIGIDITY produced by rapid motion, 215, 216
Quaternions, 135
Quicklime, 72
Quick-moving shafts, How to diminish friction in, 14

RADIAN, Definition of, 235
Radius of Curvature, 223
— — — of any fibre in bent beam, how found, 108
— — — — elastic curve, 110
— — gyration, 144
— — — of pendulum, 196
Ratio, Velocity, 6, 27
Reaction of- fluids, Application of the principle of, 209
Reciprocal figures, 148
Reciprocating mass in locomotive engine, 220
Recoil of gun, 209
Red pine, or Memel timber, 77
Re-entrant edges or corners, weak points, 82, 99
Relation between bending and twisting of cylindric shaft, 112

Resilience of cylindric spiral springs, 176
Resultant, Definition of, 136, 228
— force on joint of arch ring, 129, 163
— of a number of forces, how found, 137, 139
Rib, Arched, is in unstable equilibrium, 161
Rigid bodies, Meaning of the term, 103
Rigidity, Modulus of, N, 87, 90, 92
— produced by rapid motion, 215
Riveted joint may break in several ways, 88
Rivet holes usually weaken the metal, 88
Rocks, granitic, History and character of, 71
— stratified, History and character of, 71
Roof, Necessity for detail drawings of, 155
— Weight of snow on, 153
— Wind pressure on, 154
Roof - principal, Investigation of stresses in, 150—153
Rotating body, Energy stored up in, 42—51
Rotational periodic motion, 192
Round, Meaning of the term, 236
Rupert's drop, Condition of the glass in, 80
Rupture produced by shear stress, 88, 89, 96

SAFETY, Factor of, 120
— — — Table giving usual values of, 121
— valve, Weighted, 24
Sandstones, Character of, 71

INDEX.

Screw, Mechanical advantage of, 18
— Pitch of, 18
— propeller of vessel, Exercise on, 19
Section of beam varied for uniform strength, 119
— — loaded beam or arch, resultant of stresses at, 128—130
— — structures, stress at, how calculated, 157, 158
Set, Permanent, 55, 100
Shafts, cylindric, Relation between bending and twisting in, 112
— — Strength of, 96
— effects of twisting couple on different sections, 97—99
— Effects produced by wheels fixed eccentrically on, 218, 219
— Friction at bearings of, 12
— — in quick-moving, 14
— Power transmitted by, how measured, 31—35
— Practical rule for strength of, 95
— Stiffness necessary as well as strength, 100
— Torsional vibration of, when transmitting power, 95, 101
Shape of a loaded beam, 145, 146
— — — — chain, 160
— — wheel teeth, 27, 28
Shearing and punching, 89, 89
— force in beams and girders, 104, 114, 115
Shear strain, 86—101
— — Nature of, 90
— — pure, How to produce, 86, 89
— — and shear stress, Relation between, 92
— stress, 86—101
Simple pendulum, 39, 184, 187, 194
Sine of an angle, 235
Sines, Curve of, how drawn, 189, 200, 204

Sines, Curve of, interpreted, 200
— — — traced by a pendulum bob, 201
Skeleton drawings instructive and necessary, 28—30
Slide and piston-rod, 13, 14, 220
Slide valve, Motion of, 30
— rule, Use of, 136
Solids, Laws of friction between, 251
Specific gravity, Definition and examples of, 256
— — table of values, 68, 69
Spherical boiler, Strength of, 61
Spinning bodies, precessional motion of, 215
Spiral spring, cylindric, Angular motion of, due to turning moment, 178
— — — Behaviour of, when weighted, 178, 183
— — — Elongation in different cases, 176
— — — in dynamometer coupling, 31, 35
— — — Investigation of the forces in a weighted, 173
— — — Experiment showing relation between extension of spring and twisting of wire, 174
— — — Resilience of, 176
— — — Strength of, 175
— — — Twisting moments of, experimentally determined, 197
— — — Ultimate elongation, 177
— — — Work stored up in, 176
— — Flat, Angle of winding proportional to couple in, 170
— — — Bending moment in, 168
— — — Change of curvature of, 167
— — — Curve for turning moment of, 171
— — — Isochronism, how usually obtained, 170

Spiral spring, Flat, Turning moment proportional to winding, 171
Squared paper, Use of, 7—10, 15, 55
State of strain affects elastic strength, 62, 63, 99, 193
Statical friction, 246
Statics, Graphical, 135—166
Steadiness of Machines, 48, 219
Steam-engine, Condensing and non-condensing, 50
— — Fly-wheel for, 50
— — giving out energy, 26
— — indicated horse-power, how found, 233
Steel, Carbon and impurities in, 84
— Strength of, 84
— Tempering of, 66
Stiff joints in structures, 156
Stiffness important in shafting, 100, 101
— of beams, how it varies with linear dimensions, 125, 126
Stilling of vibrations, 200
Stone, 71
— An artificial, how made, 72
— Preservation of, 71, 72
Storage of energy during impact, 212
Store of energy in a moving body, 39, 75
Straight line figures, Properties of, 146—148
— — on squared paper, Meaning of, 9, 10
— — what it may represent, 135
Strain, 51—58
— energy, Storage and transmission of, 64, 65
— in bent beam, proportional to distance from neutral surface, 102
— Nature of, in a wire, 56, 57
— proportional to stress, 53, 55

Strains due to contraction in cooling, 79—83.
Stratified rocks, 71
Strength and stiffness of beams, Tables giving, 116—118, 121
— — best section of beam for, 106
— Elastic, 59, 99—100
— modulus for beams of different sections, 106
— of rectangular beams, how it varies with breadth and depth, 121, 125
— — — — supported at ends and loaded in middle, Table, 121
— of columns, long and short, 130
— — — Gordon's rule for, 131—133
— — cylindric spiral spring, 175
— — flat plates, 133, 134
— — pipes and boilers, 59—61
— — riveted joints in boiler plates, 88
— — shafts, 95—101
— — similar structures similarly loaded, 134
— — structures with stiff . joints, 156
— — teeth of wheels, 133
— — timber, 77, 78, 121
— Table of, for different materials, 68, 69
Stress, 52—61
— Amount of, anywhere in a section, 106, 107
— Breaking, for beams, 120
— — — columns or struts, 131
— — table of values, 68, 69
— diagrams, Examples of, 155
— how represented, 135
— tensile and compressive often repeated, Effect of, 65
— working shear, 88
Stresses at joints of structures, Methods of finding, 157, 158

INDEX. 269

Stresses in a hinged structure, Conditions for calculating, 149
— — loaded chain of suspension bridge, 160
— — — links, 159
— — roof - principal, how determined, 150—153
Strip of steel, Forms assumed by, 109
— — — Time of vibration of, 186
— — — Vibration of, in two directions, 192
Structures, Hinged, 146—149
— similar and similarly loaded, Strength of, 134
Struts, 53, 54, 150
— Gordon's rules for strength of, 131—133
— long, Modes of breaking, 130
— with ends fixed, and hinged, 131
Suddenly applied load, Effect of, 63, 64
Supporting and loading, Effects of different methods of, 113, 116—122
Suspension, Axis of, 195, 217
— Bifilar, 200
— bridge, Stresses in chain of, 160
Symbols, Algebraic, 1

TACKLE, Blocks and, mechanical advantage, 17
Teak timber, 78
Teeth of wheels, Cycloidal, 28
— — — like beams fixed at one end. 133
— — — Pitch and number of, 27
— — — — Rule for, 133
— — — Pressure on, 133
— — — Shapes of, 27, 28

Teeth of wheels, Strength of, 133
Temperature, Effects of, on beams and supports, 113
Tempering steel, Method of, 66
Tensile strength of materials, 58, 68, 69, 134
Thrust at crown of arch, 163, 165
Tie-rod, 53, 150
Timber, Ash, 78
— beams, Strength of, 120, 121
— Beech, 78
— Cedar, 78
— Durability of, 78
— Elm, 78
— Felling of, 78
— Larch, 77
— Mahogany, 78
— Memel, 77
— Oak, 78
— Preservation of, 79
— Seasoning of, 79
— Teak, 78
— Warping of, 77
— White fir or Norway spruce, 77
Time of vibration of balance of a watch, 193
— — — — bar hung at end of a wire, 193
— — — — compound pendulum, 195
— — — — liquid in bent tube, 187
— — — — magnet, 200
— — — — simple pendulum, 185
— — — — spiral spring, 183
— Periodic, of a pure harmonic motion, Rule for, 183
Torque, Definition of, 231
Torsional vibrations of shaft and want of stiffness, 100, 101
Torsion of shafts and beams, 93—101
Transmission dynamometer, 35, 36
— of power by belts and pulleys, 32

Transmission of power by shafts, 30, 95—101
— — — — wheels, 27, 28, 133
— — strain energy, 64, 65, 207, 211—214
Triangle of forces, 3, 21, 229
Twist, Angle of, 31, 93
— how measured, 30, 31, 93—96
Twisting and bending combined in shafting, 100
— — — Relation between, 112
— moment, 93—101
— — Experimental investigation of, 197—199

U TUBE, Vibration of liquid in, 186, 187
Unclosed polygon, 137
Uniform velocity ratio given by cycloidal teeth, 28
— strength in beam, 119
Use of elementary principles, 70

VALVE, Safety, Example of, 24
— slide, Motion of, 30
Variable things compared, 7, 10
Velocity, Definition of, 224
— Angular, 235
— at any point in pure harmonic motion, how found, 180
— of bullet, how measured, 218
— ratio, 6, 27
Vents, Use of, in moulding, 81
Vertical line, Definition of, 223
Vibration, Amplitude of a, 201
— Nature of a, 40, 187
— Time of, for simple pendulum, 185
— Torsional, of a shaft, 101

Vibrations, Damped, Law for, 201
— Investigation of damping of, 201
— Representation of, 204
— Stilling of, 200
Viscosity measured by twist in wire, 199, 248
— Relative, Example worked out, 205
Voussoirs, Pressure on, 162

WALL, Pressure of earth against, 73, 74
Watch balance, Motion of, 193
Waterfall, Energy in, 6
Water flowing through an orifice, Velocity of, 76, 77
— Friction of, in pipes and pumps, 75, 76
— Pressure of still, 74
— pressure on any surface submerged, 74, 75
— Quantity of, flowing through a pipe, how calculated, 77
— Total energy of, 75
Waterpipes, Strength of, 59, 60
Weighbridge, Mechanical advantage of, 24
Weight of a body, 223, 224
Wheel and axle, Mechanical advantage of, 20
— Necessity for drawings of, 28
— of locomotive, Balance weights on, 221
— teeth, Pressure on, 133
— — Shapes of, 27
White fir or Norway spruce, 77
Wind pressure against roofs, 154
Work, Definition of, 232
— how measured, 5, 6, 15, 16, 31, 32, 232, 233

Work, Law of, 5, 17, 231
— lost in friction, 11—17
— Rate of doing, 15, 233
Working stress, 59
Worm and worm-wheel, 28
Wrought iron, 83, 84

YOUNG'S modulus of elasticity, E, 54, 69, 92

ZINC, Alloys of, with copper, 85

THE END.

www.ingramcontent.com/pod-product-compliance
Lightning Source LLC
Chambersburg PA
CBHW031341230426
43670CB00006B/405